T5-BYG-710

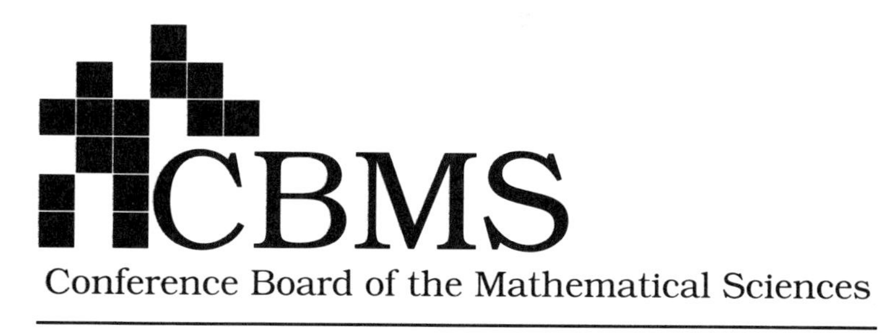

Issues in Mathematics Education

Volume 3

Mathematicians and Education Reform
1990–1991

Naomi D. Fisher
Harvey B. Keynes
Philip D. Wagreich
Editors

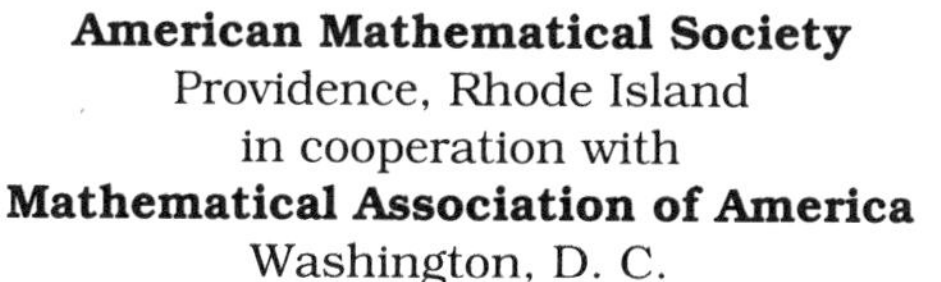
American Mathematical Society
Providence, Rhode Island
in cooperation with
Mathematical Association of America
Washington, D. C.

This volume was compiled by the Mathematicians and Education Reform Network (MERNetwork). Its activities for 1990–1991 are supported by National Science Foundation Grant TPE-8850359.

1991 *Mathematics Subject Classification*. Primary 00A99;
Secondary 00A35.

ISBN 0-8218-3503-3
ISSN 1047-398X

Copying and reprinting. Individual readers of this publication, and nonprofit libraries acting for them, are permitted to make fair use of the material, such as to copy an article for use in teaching or research. Permission is granted to quote brief passages from this publication in reviews, provided the customary acknowledgment of the source is given.

Republication, systematic copying, or multiple reproduction of any material in this publication (including abstracts) is permitted only under license from the American Mathematical Society. Requests for such permission should be addressed to the Manager of Editorial Services, American Mathematical Society, P.O. Box 6248, Providence, Rhode Island 02940-6248.

The appearance of the code on the first page of an article in this book indicates the copyright owner's consent for copying beyond that permitted by Sections 107 or 108 of the U.S. Copyright Law, provided that the fee of $1.00 plus $.25 per page for each copy be paid directly to the Copyright Clearance Center, Inc., 27 Congress Street, Salem, Massachusetts 01970. This consent does not extend to other kinds of copying, such as copying for general distribution, for advertising or promotional purposes, for creating new collective works, or for resale.

Copyright ©1993 by the American Mathematical Society. All rights reserved.
The American Mathematical Society retains all rights except those granted
to the United States Government.

Printed in the United States of America.
The paper used in this book is acid-free and falls within the guidelines
established to ensure permanence and durability. ♾

This volume was typeset using $\mathcal{AMS}$-TEX,
the American Mathematical Society's TEX macro system.

10 9 8 7 6 5 4 3 2 1   98 97 96 95 94 93

157228

Belmont University Library

QA 11 .A1 M277 1993

AAM-0918

Contents

Foreword

When we begin compiling a volume of *Mathematicians and Education Reform*, we review articles as stand-alone pieces; then as the collection takes shape we can recognize common themes among the articles. This mingling of themes and ideas between and among the articles delineates further the discussions within the articles and gives additional meaning to the collection as a whole.

In many cases, the issues and projects discussed in this volume are directed to elevating the achievement level of the general student, i.e., those students who are not necessarily going to pursue exclusive paths in mathematics and science. Think of this as the concern to improve the mathematical lot of the student-at-large. From this viewpoint, it becomes very important to gain an understanding of these students, who they are, where they are, and how to reach them. Students' experiences outside the classroom, whether from years past or in the present, influence the success or failure of the mathematics learning in the classroom. In their discussions about students, both Keynes and Roberts present a variety of influences on students to be reckoned with in mathematics education; Edwards looks at the effects of student demographics, student study habits, and student lifestyles on reform efforts; and Mahoney details a comprehensive program that reaches into grade school to identify students with academic promise and to cultivate their academic interest and development until they enter college.

Another issue, classroom dynamics or the mathematical environment, i.e., the format and content of mathematical instruction, is given careful attention. Although not always posed explicitly, two questions are being addressed: "How can we make students more responsive to mathematics?" and "How can we enable students to take charge of their own learning in mathematics?" Ideally, learning mathematics is a participatory activity with students taking more and more responsibility for exploring new directions. The power of learning environments in which students communicate with one another and in which students formulate questions is brought out in varied settings. Some examples, such as those of Davis, Fellows, and Gilman, conjure up pictures of laboratory-style classrooms alive with activity. But even modest changes, Maher emphasizes, can be very beneficial.

Taking advantage of the inextricable relationship between the process of learning and the content of the curriculum can result in an appreciably richer mathematical context than is usually seen. For example, in introducing their work on computer animation to teach mathematics, Apostol and Blinn first focus on "the power of visual images to stir [our] deepest emotions", and Kasimatis and Sallee emphasize that their work in developing a new college preparatory mathematics course is distinguished more for "*how* the material is presented...than *what* is presented".

Any thoughtful discussion about reforming mathematics education must include the professional development of teachers, as Keynes, among others, points out. In three different teacher-education projects, Braunfeld, Moussa and Goldman, and Haimo each develop teachers' confidence in their knowledge of mathematics as the heart of good mathematics teaching. Casey, drawing on the analogy of the whole language movement, suggests that professional development is more than accruing knowledge; teachers first need to see themselves as mathematicians in order to convey to their students the fullness of doing mathematics.

Naomi D. Fisher
University of Illinois at Chicago

Projects

CBMS Issues in Mathematics Education
Volume 3, 1993

The Ohio State University Young Scholars Program

CAROLYN R. MAHONEY

INTRODUCTION

The Young Scholars Program[1] (YSP) was initiated in 1988 at the Ohio State University (OSU) as part of that institution's commitment to the reversal of the disturbing trend of decreasing minority enrollment in higher education nationwide. Although the number of African-American and Hispanic young adults is growing at a rapid rate, relatively few are making their way to college and even fewer successfully complete college. The Education Commission of the States (ECS) reports that by the year 2020, Caucasion children will represent only one of every two students in our nation's public schools, yet in 1988 eight out of every ten students in higher education was white, and in 1987 minority students received only 12% of baccalaureate degrees awarded [**16**, **13**]. The YSP is a statewide effort to increase the number of African-American, Hispanic, Appalachian, and other underrepresented students who are prepared to succeed in degree programs at four-year institutions of higher learning. This article describes the mathematics component of this comprehensive program.

DESCRIPTION OF PROGRAM

Each spring, teachers, principals, and guidance counselors nominate sixth grade students from specific low-income, urban areas in Ohio. Nominations may also come from parents and from students who wish to become Young Scholars. Students are selected on the basis of academic performance, test scores, leadership, community involvement, and family circumstances. All of the students are from low-income families and are members of underrepresented groups. Most will be the first in their family to graduate from college.

[1]This program should not be confused with the NSF initiative bearing the same name.

©1993 American Mathematical Society
1047-398X/93 $1.00 + $.25 per page

Young Scholars are selected from the public schools of Akron, Canton, Cincinnati, Cleveland, Columbus, Dayton, Lorain, Toledo, and Youngstown. These nine cities contain over 75% of the minority students enrolled in the public schools of Ohio. About 93% of OSU's Young Scholars are African-American, 5% are Hispanic, and 2% are Appalachian. Public school officials, representatives from OSU, and members of the students' communities make the final selections of Young Scholars. The program admits 400 sixth graders each spring, with equal numbers of males and females. Young Scholars and their parents must make a commitment to participate actively in all elements of the program. Scholars are expected to enroll in college preparatory courses and to maintain satisfactory grades.

Young Scholars are given year-round educational opportunities and personal support before they graduate from high school. During their six years in the program, Young Scholars will participate in two-week or three-week institutes every summer on the OSU campus. The institutes include academic classes and workshops in career exploration, study skills, and personal development. Young Scholars continue to strengthen their academic preparation in their home communities through the support of teachers and mentors. Academic enrichment activities are provided year-round. YSP staff work with principals and teachers to assure that all is proceeding well. Program staff work with parents to show them how to support their children in educational pursuits, study skills, and learning about career opportunities. Parents also attend cultural events with their children. Young Scholars are each matched with a college educated professional from the same city who serves as a mentor and role model. Mentors and Scholars spend several hours each month discussing school, career, extracurricular, and personal activities, and attending events of interest to both. Tutors are available during the summer institutes and the school year [**14**].

The pre-seventh grade summer institute has three academic components: Biological Sciences, English and Ethnic Studies, and Calculator Enhanced Pre-Algebra. The pre-eighth graders take The Influence of Africa: Music and Movement in Our Culture, Critical Thinking, and Spatial Visualization. The pre-ninth grade scholars enjoy Unified Science, The American Experience Through Literature, and Probability. Additional activities include computer workshops, health education, and career exploration. These first three summers, the students are in residence for two weeks; the final three summer institutes are of three-week duration. During the institutes, Young Scholars live and eat in residence halls on the OSU campus, along with their counselors and some of the program staff. They have access to the same recreational facilities, medical services, and libraries that are available to regular OSU students. Dr. James Bishop, Special Assistant to the Provost, oversees all aspects of this program.

Mathematics Component of the Young Scholars Program

As indicated above, the mathematics component of the pre-seventh grade summer institute consists of Calculator Enhanced Pre-Algebra, the pre-eighth grade institute features Spatial Visualization, and the pre-ninth grade institute offers Probability. The mathematics curricular plan for the remaining years consists of Graph Theory, Number Theory, and an integrated mathematics-science capstone experience. Algebra classes meet for 60 minutes, while Spatial Visualization and Probability each meet for 75 minutes. The teaching assistants are OSU undergraduate, graduate, or professional students. The mathematics teachers are regular teachers of middle school, high school, junior college, or university, with ethnic backgrounds including African-American, Hispanic, and Caucasian.[2] The university mathematicians who have been involved in the summer institutes to date include: Franklin Demana, Professor of Mathematics at the Ohio State University; Carolyn Mahoney, Professor of Mathematics at California State University, San Marcos; J. Phillip Huneke, Professor of Mathematics at the Ohio State University; and Joseph Fiedler, Assistant Professor of Mathematics at California State University, Bakersfield.

In 1988, there was one summer institute designed to serve 200 pre-seventh graders. The 1989 Program was designed to serve the returning students (now pre-eighth graders) and 400 new pre-seventh graders; this service was accomplished in three sessions, each of approximate size 200. The 1990 Institute hosted 1,000 students in three sessions composed as follows: 400 rising seventh graders, 400 rising eighth graders, 200 rising ninth graders.

A one-week training session is held one week before any students arrive. This training includes overall program orientation and academic component training. On two days of this week, YSP office staff, component coordinators, teachers, teaching assistants, and dorm counselors meet together to learn the history, purpose, and philosophy of the program and to learn and discuss ways to best understand, motivate, instruct, and inspire the student population. On the other days, each component gathers its staff together to begin training. In mathematics, this training continues the discussion about program philosophy and goals using such references as *Everybody Counts* **[10]** and the *National Council of Teachers of Mathematics (NCTM) Standards* **[12]**. During this week and throughout the two-week institute, there are daily mathematics staff meetings wherein teachers take turns presenting lessons and leading the group in a discussion of how best to convey the material. There is a healthy exchange of ideas on underlying mathematical content and goals,

---

[2] In 1988, there were four teachers for the one session: all female (F), one Hispanic (H), two African-American (B), one Caucasian (C). There was one elementary school teacher (E), one high school teacher (S), one junior college teacher (J), and one university professor (U). In 1989, there were six teachers for the three sessions: 3BF, 1CF, 1CM, 1HF; 1E, 3S, 2U. In 1990, there were 12 teachers for the three sessions: 4BF, 1CF, 5CM, 2HF; 1E, 8S, 3U.

expected student response, classroom management techniques, etc. These daily meetings have proven to be a crucial factor in the overall success of the program.

Algebra. Percent was chosen as a primary topic for the first summer because it is traditionally thought to be a difficult topic for minorities, and because it is rich in realistic and interesting applications. The textual materials used are the supplementary materials **[6]** developed in the Approaching Algebra Numerically (AAN) project funded by BP America and by the National Science Foundation. The goal of these materials is to establish the concepts of variable and function before the students encounter the formalities of algebra. The materials place a heavy emphasis on problem-solving and are significantly dependent on the use of hand-held calculators. The guess and check approach afforded by the use of calculators facilitates investigating, exploring, and mathematical talk about concepts **[9]**. Having the students create, in writing, a real-world problem which can be solved by a given equation helps the assimilation process. The goal for the two weeks is to build as much student understanding as possible about markups, discounts, and population changes over time.

While it is difficult to generalize about student performance based on two weeks of instructional activity, we present here some data which suggests short-term success. The same test was given as a pre-test and a post-test to the 200 summer 1988 YSP participants. The test items consisted of typical questions from the GRA materials. Three of the items appeared on the pre- and the post-tests used to evaluate the progress of the seventh grade students in the AAN project. One of the three YSP items also appeared on the *Second International Mathematics Study* (*SIMS*) **[17]** test which measured pre- and post-test performance of United States eighth grade students as well as those of participating foreign countries. Detail about these three items is given below. The YSP students scored 23% on the pre-test and 59% on the post-test. Interested readers can obtain a copy of the pre- and post-tests by writing the author.

1. The first item considered is:

YSP test:	Write 0.28 in percent form
AAN test:	Write 0.72 in percent form

We view these two items as comparable. The mean correct score of the AAN project seventh graders, the AAN project control seventh graders, and the YSP soon-to-be-seventh graders on the items was as follows:
It is interesting to note that the AAN students took the pre-test at the beginning of the seventh grade and the post-test at the end of the seventh grade. The YSP students had not yet entered the seventh grade.

2. The second item considered is:

	Pre-test	Post-test
AAN project students	58%	88%
AAN control students	54%	82%
YSP students	53%	89%

YSP test:	Determine the number of free throws tried if a player made 56 free throws which represented 80% of the number tried.
AAN test:	30 is 75% of what number?

Many would argue that the two items are not really equivalent. One reason comes from the growing body of evidence that suggests problems based in reality are somewhat easier than problems out of context. This point of view suggests that the AAN test item is a little harder. However, the YSP item appeared in a table and the students had to first realize that the question was about determining what number 56 was 80% of. The AAN test item also appeared as a core item on the SIMS test, so we not only have the United States eighth grade data but an international post-test median. The results on these two items were as follows:

	Pre-test	Post-test
AAN project students	48%	71%
AAN control students	44%	54%
SIMS US eighth graders	44%	51%
SIMS International	–%	48%
YSP students	53%	89%

The fact that the YSP students took the post-test so soon after to the pre-test might account for some of the difference in the performance of the YSP students as compared to the United States eighth graders.

3. The last item considered is:

YSP test: The population of Nelsonville is increasing at the rate of 2.5% each year. Complete Table 1.

TABLE 1. The population of Nelsonville.

Year	**Population**
1982	80,000
1983	
1984	
1985	
n years after 1982	

AAN test: The population of Clarktown is 12,480. If the population increased by 3% next year, what will its population be?

The results for the AAN test item and the first row to be completed in the YSP item are:

	Pre-test	Post-test
AAN project students	37%	75%
AAN control students	29%	35%
YSP students	6%	39%

The supplementary materials used in the AAN project consisted of four chapters, each requiring three to four weeks of instructional time. The four chapters were spread over the school year. This explains why the AAN students' performance was much stronger on this item. However, with only two weeks of instruction, the post-test performance of the YSP students was also most impressive.

Spatial visualization. The need for early exposure to three-dimensional geometry and to the visualization of such using two-dimensional drawings is well documented; for example, the *NCTM Standards* calls for its inclusion in grades 5–8. In conjunction with the text, *Spatial Visualization* from the Middle Grades Mathematics Project **[18]**, concrete manipulatives are used to help students make the transition from concrete to abstract thinking.

Students work in groups as they construct buildings from mat plans, draw plans of buildings they have constructed, and practice mirror drawings. Individual roles and responsibilities in the production of group work is discussed. Group discovery also allows students to internalize the experience of research and applied mathematicians.

Conversation and coordination among and between the various program components is a major goal of the program staff. For example, in this spatial visualization component, there are lessons which require the students to produce mirror images of given drawings. After the first of these lessons, the students visit the dental school as one of their career exploration outings. The next lesson includes conversation about their experience with the dental mirrors.

In the spatial visualization component, a test is given at the end of each week. After two years of administering the multiple choice test which accompanies the text, the staff designed a free response assessment instrument. In 1991, the program studied individual and group performance on spatial visualization by a group of eighth grade students, including some of the Young Scholars.

Probability. The citizen of today is inundated with quantitative information involving charts, graphs, percentages, and averages—information that has great impact on health, citizenship, parenthood, jobs, and other important matters. The ability to deal with data using logical reasoning and taking advantage of current technology is required of all well-functioning citizens. A survey of the Young Scholars revealed that many of the students had not been sufficiently exposed to probability in the elementary grades. Hence, in the summer of 1990, the first class of approximately 200 pre-ninth graders immersed themselves in probability as presented in the Middle Grades Mathematics Project text *Probability* [**15**]. The unit reinforced the concept of fractions introduced in the pre-seventh grade component, exposed students to charts and histograms, and taught them to use spreadsheets. While the students used computers for word processing, to enhance and reinforce the spatial visualization component of their previous summer, and for recreation, this was the first year computers were an integral part of their mathematics class. Working in groups, the Young Scholars conducted experiments, made conjectures, and tested their conjectures. Students learned to make decisions, both mathematical and informal, as they used blocks, marbles, balls, play money, etc., to design experiments and to draw conclusions.

Academic Year Enrichment

The YSP presents weekend academic enrichment activities for participants at monthly intervals during the school year. The events are generally three hours long and held on a Saturday at a local school or college, or at a government or business site. Local teachers and mathematicians, in consultation with the YSP staff, are responsible for the mathematics portions of the sessions. For example, Dr. Sherwood Silliman, Professor of Mathematics at Cleveland State University, led sessions at Cleveland State using the *Stella Octangula* materials developed by the Visual Geometry Project at Swarthmore, and Dr. Karl Maneri, Professor of Mathematics at Wright State University, led numerous sessions at several Dayton area institutions. The programs are intended to sustain enthusiasm and to reward the perseverance of the program participants of the summer institute. The summer material is not repeated, but the exploring, problem-solving strategy and approach to mathematics is continued. Career exploration is incorporated into some of the sessions, led by professionals in business or industry who use mathematics in their careers.

These events also allow YSP staff to meet with the students and sometimes their parents and their teachers to update one another regarding the students' current progress in school. During the first Saturday program of the academic year, a local school mathematics supervisor talks with parents about what the students will be doing in school mathematics, how parents can help the students, and about the role of mathematics in future studies and careers. A YSP staff member explains the YSP program to local teachers and discusses ways they can become involved in the program. Ensuing sessions are designed for students only; however, some parents attend to help out or observe. Local classroom teachers are employed to facilitate and assist the university professor who leads the session. The final session of the year brings everyone back together to help evaluate the effectiveness of the enrichment activities.

Academic year enrichment programs were begun in 1989–90 for the Cleveland, Columbus, and Dayton school systems from an Eisenhower grant from the Ohio Board of Regents. Resources from the National Science Foundation allowed three more school districts, Lorain, Cincinnati, and Akron, to enjoy 1990–91 enrichment activities and all eight school districts to participate in the 1991–92 activities.

Conclusion

The most important measure of the success of the Young Scholars Program is the successful completion of college by its participants; hence, the program plans long-term monitoring of the participants. Presently, the program staff records grades, attendance, and demographic data on the Young Scholars and their alternates. Short-term progress is assessed by each academic component during each Institute. In mathematics, staff are concerned about the Young Scholar's attitude toward mathematics, his or her performance in the local school environment, and the mathematics learned during the Summer Institute. Early indications are that the students' attitudes toward mathematics and their confidence in their ability to do mathematics, as assessed by questionnaires, are improving. In an autumn 1990 proficiency test administered to ninth graders throughout Ohio, only 17% of African-American students passed the mathematics section. Early, but incomplete, data indicate that more than two-thirds of the Young Scholars passed the mathematics section of this proficiency exam.

In the Pre-Algebra Institute, the same test is administered as a pre-test and a post-test. In Geometry and Probability, students are tested at the end of each week. Daily homework is graded and promptly returned and discussed. The teacher-student ratio of 2:17 allows for much interaction between academic staff and students; such interaction, together with daily staff meetings, allow formative evaluation and monitoring of students' progress. Additionally, at the end of each week, the entire academic staff (all components) meets to evaluate the preceding one or two weeks and to suggest changes and/or improvements.

This article has outlined the background and history of the OSU's Young Scholars Program. Future articles will more carefully assess successes and failures.

REFERENCES

1. Anne Chapman, *The difference it makes: A resource book on gender for educators*, National Association of Independent Schools, 1988.

2. Reginald M. Clark, *Critical factors in why disadvantaged students succeed or fail in school*, Academy for Educational Development Inc., 1988.

3. Beatriz Chu Clewell, Margaret E. Thorpe, and Bernice Taylor Anderson, *Intervention programs in mathematics, science, and computer science for minority and female students in grades four through eight*, Educational Testing Service, Princeton, NJ, 1987.

4. Commission on Minority Participation in Education and American Life, *One third of a nation*, American Council on Education and Education Commission of the States, Washington, D.C., 1988.

5. Barbara Davis and Sheila Humphries, *Evaluating intervention programs: Applications from women's programs in math and science*, Teachers College Press, New York, NY, 1985.

6. Franklin Demana, Joan R. Leitzel, and Alan Osborne, *Level 1—Getting ready for algebra*, D. C. Heath and Company, Lexington, MA, 1988.

7. Elizabeth Fennema and Gilah Leder, editors, *Mathematics and gender*, Teachers College Press, New York, NY, 1990.

8. Hudson Institute, *Workforce 2000*, Indianapolis, Indiana, 1987.

9. Carolyn R. Mahoney and Franklin Demana, *Filling the math and science pipeline with young scholars*, Notices Amer. Math. Soc., **38**, no. 2 (1991), 101–103.

10. Mathematical Sciences Education Board National Research Council, *Everybody counts: A report to the nation on the future of mathematics education*, National Academy Press, Washington, D.C., 1989.

11. Mathematical Sciences Education Board National Research Council, *Reshaping School Mathematics: A Philosophy and Framework for Curriculum*, National Academy Press, Washington, D. C., 1990.

12. National Council of Teachers of Mathematics, *Professional standards for teaching mathematics*, Reston, VA, 1991.

13. National Task Force for Minority Achievement in Higher Education, *Achieving campus diversity: Policies for change*, Education Commission of the States, Denver, Colorado, 1990.

14. The Ohio State University, *Young scholars program* (brochure).

15. Elizabeth Phillips, Glenda Lappan, Mary Jean Winter, and William Fitzgerald, *Probability*, Middle Grades Mathematics Project, Addison-Wesley Publishing Company, Menlo Park, CA, 1986.

16. Richard C. Richardson, Jr., *Promoting fair college outcomes: Learning from the experiences of the past decade*, Education Commission of the States, Denver, CO, 1991.

17. Kenneth J. Travers, Director, *Second international mathematics study, Detailed national report United States*, Stripes Publishing Company, Champaign, IL, 1985.

18. Mary Jean Winter, Glenda Lappan, Elizabeth Phillips, and William Fitzgerald, *Spatial visualization*, Middle Grades Mathematics Project, Addison-Wesley Publishing Company, Menlo Park, CA, 1986.

DEPARTMENT OF MATHEMATICS, CALIFORNIA STATE UNIVERSITY, SAN MARCOS, CALIFORNIA
E-mail address: c-mahoney@csusm.edu

CBMS Issues in Mathematics Education
Volume **3**, 1993

Using Computer Animation to Teach Mathematics

TOM M. APOSTOL AND JAMES F. BLINN

ABSTRACT. Visualization—the visual representation of mathematical ideas, principles or problems—has always played an important role in both teaching and learning mathematics. Visualization is even more effective when the images are in motion. This article describes some of the visualization techniques employed in the first four computer animated videotapes produced by *Project MATHEMATICS!*.

1. COMPUTER ANIMATION AS AN INSTRUCTIONAL MEDIUM

The power of visual images to stir the deepest emotions has always been understood by artists. Television places these images in motion together with music and special effects. The impact of well-crafted televised images on the human mind has been exploited by entertainers, advertisers, and politicians since the advent of television.

Creative use of television for teaching science and mathematics was demonstrated in the 52-episode physics telecourse *The Mechanical Universe...and Beyond*, produced in the 1980s by the California Institute of Technology and the Southern California Consortium. The key ingredient that made this award-winning series a success was the more than seven hours of computer animation that brought mathematics and physics to life in ways that could not be done at the chalkboard or in a textbook. The animation was designed and executed by James F. Blinn (one of the authors of this article) who for many years produced planetary flyby simulations for NASA at the Jet Propulsion Laboratory.

The combination of computer animation and video provides a powerful instructional aid that: (a) grabs the viewer's visual attention and maintains the viewer's involvement, (b) capitalizes on the viewer's visual intuition, (c) portrays a large quantity and diversity of information in a brief period of time, (d) takes advantage of the viewer's sophistication in "reading" visual clues, and (e) conveys mathematics in a rich cultural context.

Key words and phrases. Audio-visual resources, computer animation, mathematics education, video technology, visualization.

©1993 American Mathematical Society
1047-398X/93 $1.00 + $.25 per page

The authors of this article, part of the same team that produced the physics telecourse mentioned above, launched *Project MATHEMATICS!* in 1987 to attract young people to mathematics through high-quality videotapes that show mathematics to be understandable and exciting. Each tape, together with a workbook/study guide, explores a basic topic in mathematics that can be easily integrated into any existing high school or community college curriculum. Computer animation is combined with live action relating mathematics to the real world and with images of original documents that weave a historical perspective.

The project was conceived under the name *Project MATHEMATICA*. After the first videotape, *The Theorem of Pythagoras*, was completed with private funding in May 1988, the name of the project was changed to *Project MATHEMATICS!* This was done to avoid confusion with commercial computer software called *Mathematica* that appeared in the summer of 1988.

By early 1992, the project had produced four additional modules with support from the National Science Foundation: *The Story of Pi, Similarity, Polynomials*, and *Sines and Cosines, Part* I. More NSF-supported modules are in production. Another module, *The Teachers Workshop*, was completed in 1991 with joint support from the Intel Foundation and the NSF. It describes a lively two-day workshop held in mid-August 1991 for teachers who had used project materials in their classrooms in a variety of settings, from grade eight in middle school to community college.

Project MATHEMATICS! videotapes and workbooks are distributed on a nonprofit basis through an extensive distribution network that includes 36 state departments of education, the Mathematical Association of America, the National Council of Teachers of Mathematics, and commercial distributors. In the academic years 1989/1990 and 1990/1991, more than 30,000 videotapes were in circulation, and at least two million students had seen one or more of the tapes. The actual numbers are undoubtedly much higher because tapes and workbooks may be copied freely for educational use, and in some states the tapes are broadcast directly to classrooms through educational networks.

Although each videotape is less than 30 minutes long, each module supports five or more periods of classroom instruction. Universally favorable response to these materials indicates that this is a successful use of video technology to aid mathematics education.

This article describes some of the visualization techniques employed in the first four tapes produced by *Project MATHEMATICS!*. The illustrations are excerpts from the computer animated sequences. A preliminary version of this article was prepared for distribution at a Chautauqua Short Course held at the California Institute of Technology, March 22–24, 1990, and was published in the March 1991 issue of *Primus*.

2. The Videotape on *Similarity*

2.1 Brief outline of the videotape on *Similarity*. After a brief computer animated *Review of Prerequisites* on ratios of numbers, the program opens by showing a variety of objects from real life having the same shape but not necessarily the same size. The narrator asks, "Can we construct a figure with the same shape as another?" A triangle is moved to various positions by translating it, rotating it, or flipping it over. They are congruent because they have not only the same shape but also the same size.

To change size without changing shape, scaling is introduced. Scaling multiplies lengths of all line segments by the same number, called the scaling factor. Scaling preserves angles; it also preserves ratios of lengths of corresponding line segments.

Applications show how Thales might have used similar triangles to find the height of a column and of a pyramid by comparing lengths of shadows. Another application explains why the sum of the angles in any triangle is a straight angle.

Similarity is then discussed for more general polygons and also for three-dimensional objects. The program then shows what happens to perimeters, areas, and volumes under scaling. Perimeters are multiplied by the scaling factor, areas by the square of the scaling factor, and volumes by the cube of the scaling factor. The abstract ideas are illustrated with a variety of examples from real life.

The concept of similarity is one of the great triumphs of Euclidean geometry, with applications extending far beyond geometry. Similarity plays an important role in every aspect of art, science, or technology involving measurement. It reveals the secret of map making, scale drawing, and blueprints, and also explains some aspects of photographic images and vision itself. Similarity also plays a role in regulating the size and structure of life forms. For every type of plant or animal there seems to be an optimum size, and similarity helps explain why a significant change in size always carries with it a change of form.

2.2 Comments on some of the visualization techniques used in *Similarity*.

2.2.1 *The Expando Ray*. In the mathematical definition of scaling, we choose a fixed point O as origin and a fixed positive number s, called the scaling factor. For every point P we associate another point P' whose distance from O is equal to s times the distance of P from O. The process of associating P' with P is called *scaling by a factor* s, and the point O is called the *center of scaling*.

One of the problems encountered in designing this program was to portray the concept of scaling in animated form. This was done by introducing a science fiction type ray gun, called the "Expando Ray". For example, to double the size of a triangle, you set the dial at 2 and aim the gun at a point inside the triangle chosen as the center of scaling. A beam of light strikes the center of scaling and lines emanate from the center to all three vertices, doubling all distances and producing a triangle twice as large. The Expando Ray aims at different points to convince the viewer that the result does not depend on the location of the center of scaling. You can aim at any point inside, on the boundary, or outside the triangle, or even outside the plane of the triangle. The expanded figure is another triangle whose edges are twice as long as those of the original triangle. The image is enhanced by sound effects that are heard when the gun is fired and when the target is struck. If the scaling factor is less than 1 the "Expando Ray" becomes a "Shrinko Ray", and when a general scaling factor is used it becomes a "Scale-o Ray". The same device is used for scaling more general plane figures and solid objects as well (Figure 1). The fact that the angles do not change under scaling is demonstrated by placing the smaller figure inside the larger and comparing the angles one at a time.

2.2.2 *Applications of similarity.* One of the oldest applications of similarity is determining the height of a tall object, such as a tree or column, which cannot be measured directly. The Greek mathematician Thales in the sixth century B.C. is said to have invented a method for determining the height of a column by comparing the length of its shadow with that of his staff. An animated segment shows that the ratio of the height h' of a column to h, the height of Thales' staff, is equal to the ratio of length w' of the column's shadow to length w of the staff's shadow. As the sun moves through the sky, the shadow lengths change but the ratio w'/w does not change. Therefore, if the shadows are measured at the time of day when the staff's shadow length w is exactly equal to the staff's height h, the corresponding shadow length w' will be equal to the column height h' (Figure 2(a)).

We showed the equality of the staff height to its shadow by having Thales drop his staff directly onto the shadow. The equation $w = h$ is presented with a touch of humor by having the symbol w bounce up and become h when the staff hits the ground. In the tradition of animated storytelling, the column also falls to the ground. As Thales realizes what is about to happen,

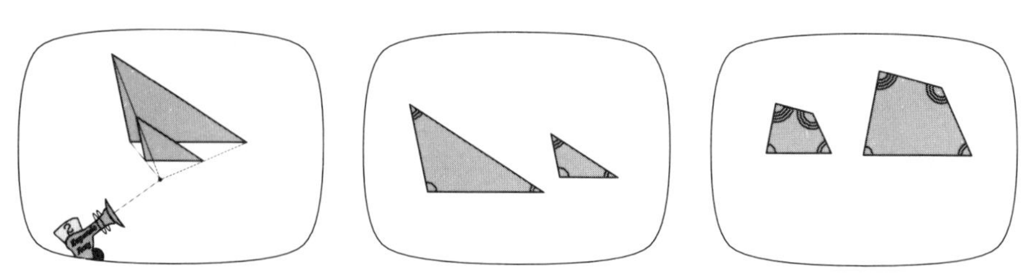

FIGURE 1. Using the Expando Ray to expand figures by a factor 2.

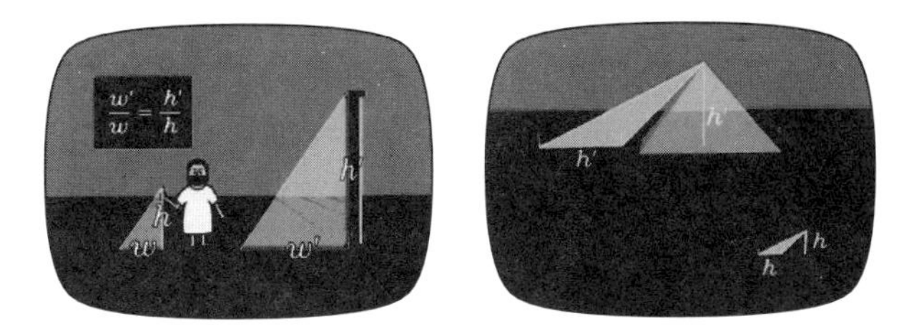

FIGURE 2. Thales determining (a) the height of a column, and (b) the height of a pyramid.

he reasons mathematically and jumps out of the way just in time. This illustrates, in a subtle way, another practical use of mathematics—survival.

Legend has it that Thales also calculated the height of a pyramid by comparing lengths of shadows. This is more of a challenge because part of the pyramid's shadow falls on the pyramid itself. An animated segment shows how this calculation can be done very simply by marking the end of the pyramid shadow when it falls outside the pyramid, and measuring the change in the shadow length at a later time of day when the staff's shadow is stretched by an amount equal to the length of the staff. The height of the pyramid is equal to the corresponding change in its shadow length (Figure 2(b)).

2.2.3 *Perimeters, areas, and volumes of similar figures.* When a plane figure is expanded or contracted by a scaling factor s, its perimeter is multiplied by s. This is demonstrated first for a triangle, then for a more general polygon, and finally for a curved figure that can be approximated by polygons.

To find the effect of scaling on the area of a plane figure, a rectangle is scaled by a factor s in the horizontal direction only, multiplying the length of its base, and hence its area, by the factor s. Then it is scaled again in the vertical direction by a factor s, multiplying its altitude and hence its area by s, so the net effect of scaling in both directions by a factor s is to multiply the area by s^2. Animation reveals that the same holds for a triangle, because the area of a triangle is one half the product of its base and altitude, and hence for more general polygonal figures made up of triangles. Finally, the result extends to figures with curved boundaries because they can be approximated by polygonal figures (Figure 3.)

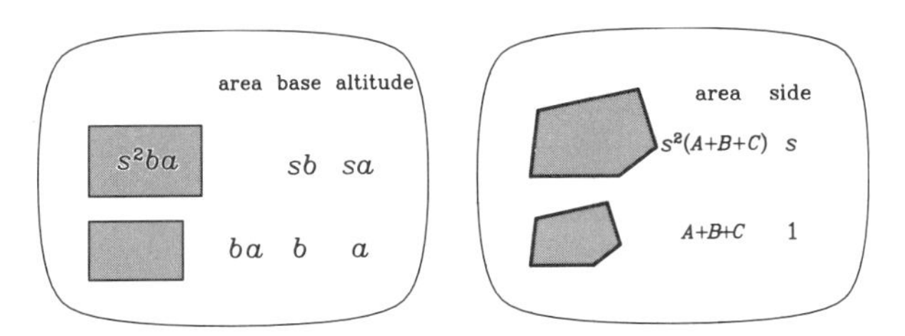

FIGURE 3. Areas of similar figures are multiplied by the square of the scaling factor.

The extension to three-dimensional solids is suggested by showing that when the edges of a rectangular box are multiplied by a scaling factor s, its volume gets multiplied by s three times, once for each dimension, so the net effect is to multiply the volume by s^3.

2.2.4 *Applications to biology.* Similarity helps explain why a hummingbird's heart beats about sixteen times faster than a human heart. A hummingbird four inches long is about one-sixteenth the average height of a person, but the amount of blood in its body is proportional to its volume, which is about one-sixteenth cubed of a person's volume. The surface area, through which body heat escapes, is about one-sixteenth squared of a person's surface area. So the bird has a surface area about one-sixteenth squared of ours, but only one-sixteenth cubed as much blood to keep it warm. Therefore, its heart has to pump about sixteen times as fast. That's roughly a thousand beats per minute.

Giants about sixty feet tall appear in classic literature such as *Jack and the Beanstalk* or *Gulliver's Travels*, and huge monsters such as King Kong or Godzilla are featured in science fiction movies. The clever use of mechanical scale models and special effects makes a movie giant like King Kong look real. But there are arguments based on similarity explaining why such creatures cannot exist in the real world, and we decided it would be a good application of similarity to include one of these arguments in the program. We attempted to obtain stock footage of giant creatures from Hollywood studios but were unable to find something we considered appropriate.

By a stroke of good luck, we learned of the existence of three minutes of experimental 16mm film made in the 1950s by Taras Kiceniuk and Robert H. Crandall, who used ingenious photographic techniques to make small live creatures such as insects appear 80 times their actual size. We acquired the only existing copy of this film and had it converted to videotape of superb quality. Because some of the creatures are shown emanating from a manhole, we called them *Manhole Monsters.* One of the monsters is a praying mantis the size of a horse, and we use similarity to explain why such a creature cannot exist in the real world. If a praying mantis is enlarged to be the size of a horse, the scaling factor is about 80, increasing its volume and hence its weight by a factor 80^3. But the strength of its legs increases like their cross sectional area, by a factor 80^2. Consequently, its legs would have to support 80 times the weight borne by a real praying mantis. The material in those legs would be crushed under this load. An entertaining animated sequence shows a giant praying mantis collapsing on those spindly legs as the narrator says, "And that shows how easy it is to destroy monsters with a little mathematics."

3. VIDEOTAPE ON *The Theorem of Pythagoras*

3.1 Brief outline of the videotape on *The Theorem of Pythagoras*. After a brief computer animated *Review of Prerequisites*, the tape begins with three real life situations that lead to problems involving right triangles. The first live action shot shows one jogger running around a rectangular track while another takes the diagonal short cut. Question: How much do you save by cutting the corner rather than going all the way around? The problem is restated with computer animated runners going along the hypotenuse and the legs of a right triangle (Figure 4).

The second live action shot shows a group of soldiers faced with the problem of how long a ladder must be to straddle a moat and reach the top of a castle wall. Animation restates the problem in terms of right triangles and adds visual interest by showing a shark swimming ominously in the moat (Figure 5).

A third live action shot shows wind prospectors looking for a good place to measure wind speed in a mountain pass. Their problem is to determine one leg of a right triangle when the other leg and the hypotenuse are known. Again, the problem is restated in animated form (Figure 6, see next page).

A brief recap shows that all three live action situations lead to the same mathematical problem:

To find the length of one side of a right triangle
if the lengths of the other sides are known.

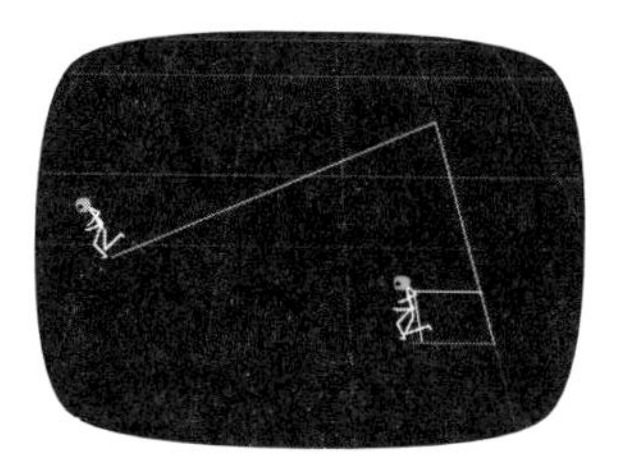

FIGURE 4. Animation relating runners to a right triangle.

FIGURE 5. Animation showing soldiers placing a ladder against a castle wall.

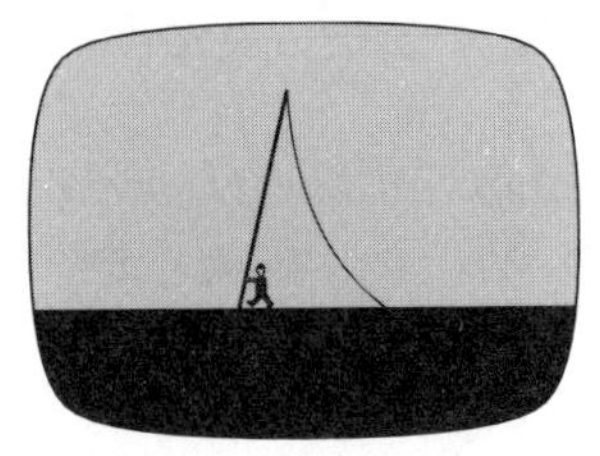

FIGURE 6. Animation showing wind prospectors erecting an anemometer tower.

The next scene solves the problem by a simple computer animated derivation of the Pythagorean theorem, based on similar triangles. The resulting algebraic formula $a^2 + b^2 = c^2$ is interpreted geometrically in terms of areas of squares and is then used to solve the three real life problems stated earlier. Historical context is provided through still images showing Pythagoras, Babylonian clay tablets, Chinese and Arabic manuscripts, and various editions of Euclid's *Elements*. The tape exhibits several different computer animated proofs of the theorem of Pythagoras, and extends it to three-dimensional space. The last segment of the tape, entitled *Previews of Things to Come*, shows how the Pythagorean theorem is used in trigonometry and points out that the theorem does not hold for spherical triangles.

3.2 Comments on some of the visualization techniques used in *The Theorem of Pythagoras*.

3.2.1 *Review of Prerequisites.* The first use of animation occurs in the *Review of Prerequisites*, which explains that the program is based on three ideas: (1) *Similar triangle ratios*, (2) *Invariance of area under shearing*, and (3) *Behavior of area under scaling.*

Similar triangle ratios. This segment reminds the viewer that lengths of corresponding sides of similar triangles have the same ratios. There are two types of ratios: internal (ratios of lengths in the same triangle) and external (ratios of lengths in different triangles). Glowing lines show the corresponding sides at the same time the ratio expressions appear on the screen. To focus attention on the object being discussed, the edges are drawn in a brighter color than the background and triangle interiors. Labels appear *on* the edges, not beside them. This is done so the labels will stay put when shapes are moved around. The labels are even brighter than the edges so they will be legible. The color scheme in this scene is chosen to match the color scheme used when this result is applied later in the program.

Invariance of area under shearing. This segment shows that shearing a triangle or a parallelogram does not change its area, a concept not often stressed in high school geometry courses. To shear a right triangle, its interior fades into a collection of horizontal line segments. The line widths and spacings are carefully chosen to be small enough to give the impression that they cover the entire interior, but large enough to be perceptible as lines. The

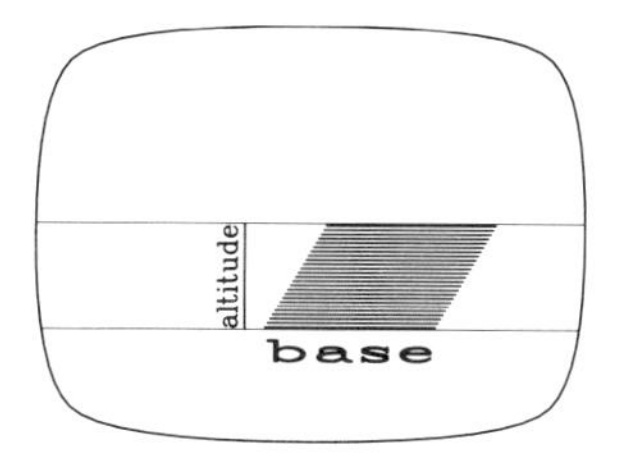

FIGURE 7. Animation showing that shearing does not change the area of a parallelogram.

base and altitude of the right triangle are labeled, and the narrator says, "The area of a triangle depends on its base and its altitude." The base is kept fixed and the triangle is sheared to form another triangle with the same altitude. As the narrator explains that the area does not change under shearing, the viewer can see this visually because the line segments covering the triangle simply move to different positions without changing their lengths. An appropriate sound effect (a sliding door) adds an audible component to the visual image and helps reinforce the concept of shearing. Incidentally, the fact that the area of a triangle is half the base times the altitude is not stated explicitly and plays no role in the visual description. This drives home the idea that invariance of area under shearing is an intrinsic property of the figure and does not depend on formulas. The scene is repeated with a rectangle sheared to form a sequence of parallelograms with the same base and altitude (Figure 7). The original rectangle fades out when its copy is being manipulated. This use of transparency deemphasizes parts of the picture that are not currently under discussion. The shearing is also done in a different direction, as the narrator says, "We can also choose another direction as base and shear in *that* direction without changing the area." This scene serves as preparation for a computer animated version of Euclid's proof of the Pythagorean theorem that occurs later in the program. Again, the color scheme in the *Review of Prerequisites* matches that used later in Euclid's proof.

Behavior of area under scaling. The scene begins with a simple example showing a unit square. One edge is highlighted. It and the area of the square are labeled 1. The square is expanded by a factor r and the narrator explains that the area of the new square is r^2. As the narrator explains that scaling a plane figure by a factor r changes its area by the factor r^2, a pentagon with one highlighted edge of length 1 appears, the label S for area is placed in its interior, and then the figure is scaled by the same factor r to produce another pentagon whose area is Sr^2.

3.2.2 *Examples from real life.*

Joggers. In the animation shown here, the labels remain static as the perspective changes and the triangle moves. The triangle is slightly transparent for two reasons. One, it shows the grid behind it to give a better feel for its size. Two, it makes the color closer to the background color to tie the shapes together visually. To emphasize that all regions in Euclidean space have common properties, patterns are not used for backgrounds. A patterned background might appear more interesting visually but would tend to blunt this perception. The two legs of the right triangle are positioned end to end to show that the sum of their lengths is greater than the length of the hypotenuse. This is intended to build intuition for the triangle inequality.

Soldiers storming a castle wall. Character animation is done very simply with rigid bodies and hinge joints at the arms and legs. The shark in the moat emphasizes that the soldiers dare not enter the moat. In the animation the ladder has the same relative size as in the live action shots. Numbers were picked that seemed about right for the scales of the objects shown.

Wind tower. Again, animated objects have the same relative size as in the live action shots. The added feature in this segment is the "twanging" sound effect as the guy wire snaps into place.

3.2.3 *Discovering the Theorem of Pythagoras.* The next significant use of animation is the algebraic derivation of the Pythagorean theorem. A perpendicular is drawn from the right angle to the hypotenuse, dividing it into two segments of lengths x and y. This divides the triangle into two smaller right triangles, each of which is similar to the original right triangle. The two small triangles are rearranged on the screen and corresponding angles are marked to show that they are, indeed, similar to the large triangle. Corresponding sides glow to emphasize equality of ratios $x/a = a/c$ and $y/b = b/c$. An "algebraic ballet" then takes place in which both sides of the first equation are multiplied by a. This is accomplished visually by having the letter a fall on each side of the equation with appropriate sound effects. As a falls on the left side, the equation tips like an unbalanced scale, and the balance is restored when a also falls on the right side. This particular animation conveys the feeling that when operations are performed on an algebraic equation, one must be careful to keep both sides of the equation equal to each other. The a is canceled on the left by an audible chalk mark, while on the right the two factors a merge to become a^2. The resulting equation is $x = a^2/c$. A corresponding ballet transforms the second equation to $y = b^2/c$ (Figure 8). The viewer is then reminded that $x + y = c$, and another algebraic ballet replaces x and y by a^2/c and b^2/c, respectively, and multiplication by c produces the formula $a^2 + b^2 = c^2$, which is the Pythagorean theorem in symbols. Incidentally, in making an algebraic substitution during an algebraic ballet, it is important to draw the viewer's attention to the symbol that is being replaced. In this case, the symbol shakes with an accompanying cowbell sound effect. Other devices used elsewhere in the series are to have the symbols glow or have them fly in from outside the screen.

3.2.4 *Geometric interpretation.* The next scene explains the geometric meaning of the algebraic formula $a^2 + b^2 = c^2$. The exponent 2 in the

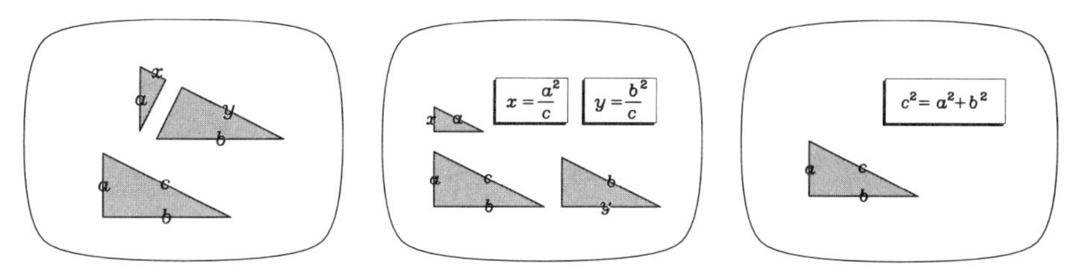

FIGURE 8. Algebraic derivation of the formula $a^2+b^2 = c^2$.

formula vibrates to call attention to the fact that we are talking about *squares* of numbers, and this suggests we interpret the formula in terms of areas of squares constructed on the sides of the triangle. The area of the square on the hypotenuse is equal to the sum of the areas of the squares on the two legs. The orientation and labeling of the scene is set up at the beginning so that the big square nudges the equation $a^2 + b^2 = c^2$ out of the way and the three geometric squares end up in the proper $a^2+b^2 = c^2$ order. This illustrates an important design consideration. Many animated scenes are designed back to front. You decide on the desired configuration for the punch line at the end of the scene, and arrange the shapes at the beginning of the scene so minimal motion is required to get there. At one time we toyed with the idea of making the three square shapes have different colors so that the larger square would represent a mixture of the two different colors of the smaller squares. But this did not work well because the color of an object (actually, its brightness) affects your perception of its size. So it was decided to use the same color for all shapes whose areas were being compared or added.

The scene is followed by a brief historical interlude showing a drawing of Pythagoras and diagrams of the Pythagorean theorem in old Chinese and Arabic manuscripts. Scenes like this are included to show that important mathematical ideas transcend national and cultural boundaries. They are produced by transferring photographic images from books or manuscripts to videotape. This is done by projecting a transparency on a special screen connected to a video camera that can zoom in to show detail or pan across the image to give the impression of motion.

3.2.5 *Applying the Theorem of Pythagoras.* The next scene shows how the Pythagorean theorem solves the three real life problems introduced at the beginning of the program. Actual numbers are inserted for two of the quantities a, b, or c in the Pythagorean equation $a^2 + b^2 = c^2$, and the formula is used to calculate the third.

3.2.6 *Pythagorean triples.* In the next animated segment, a moving ruler is shown as the hypotenuse of a variable right triangle, with one vertex sliding along a calibrated vertical ruler while another vertex slides along a calibrated horizontal ruler. Changing numbers indicate the lengths of the three sides during the motion. Since the vertical and horizontal rulers are always

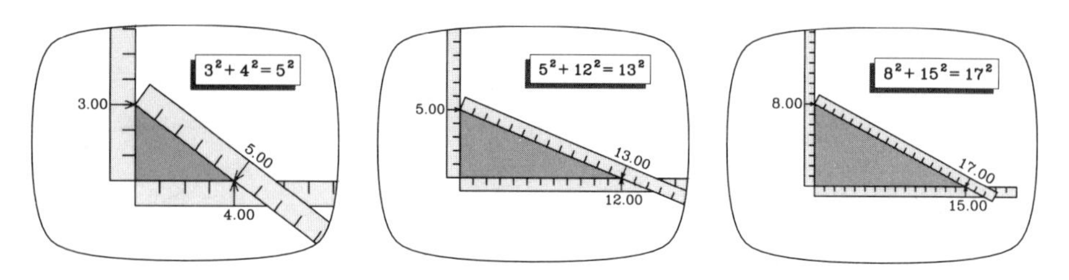

FIGURE 9. Animated segment showing examples of Pythagorean triples.

perpendicular, the three distances continuously satisfy the Pythagorean theorem. When all three sides are integers like $(3, 4, 5)$ (a Pythagorean triple), the motion stops temporarily and the Pythagorean equation $3^2 + 4^2 = 5^2$ is displayed prominently at the top of the screen. The Pythagorean triples $(5, 12, 13)$ and $(8, 15, 17)$ are also shown (Figure 9), and the narrator explains that there are many more. Students can make these discoveries for themselves by using three rulers. The color of the rulers was chosen to suggest they are made of wood.

A slide of a Babylonian clay tablet appears that contains 15 examples of Pythagorean triples inscribed in cuneiform. The corresponding section in the workbook contains several interesting exercises on Pythagorean triples accessible to high school students.

3.2.7 *Chinese proof.* The program then describes several different proofs of the Pythagorean theorem. The first is suggested by a diagram in a Chinese manuscript (Figure 10). Four copies of the original triangle, shown in yellow, are arranged around a blue square of area c^2. The yellow triangles are moved around to reveal two blue squares, one of area a^2 and one of area b^2, filling the same space as the original square of area c^2. The visual requirement here is to make sure that the blue region maintains a constant area as the yellow triangles move around. In traditional proofs, only one example of a right triangle is shown in the diagram with the implied suggestion that the same proof applies to all right triangles. Using animation, however, we can quickly show several variants of the diagram. For example, at the end of this scene, the animation is repeated with the relative sizes of the legs shown changing. This demonstrates that the final result holds for an arbitrary right triangle, including the special cases $a = 0$ and $b = 0$.

3.2.8 *Euclid's proof.* Another historical interlude explains that the Pythagorean theorem has come down to us through the ages in Euclid's *Elements.* The camera scans the pages of several visually interesting editions of Euclid's *Elements,* including a shot of the cover and a beautifully illustrated page in the first printed edition of Euclid published in Venice in 1482. Then we see the Pythagorean theorem appearing in Book I as Proposition 47. A diagram is revealed showing Euclid's proof of the Pythagorean the-

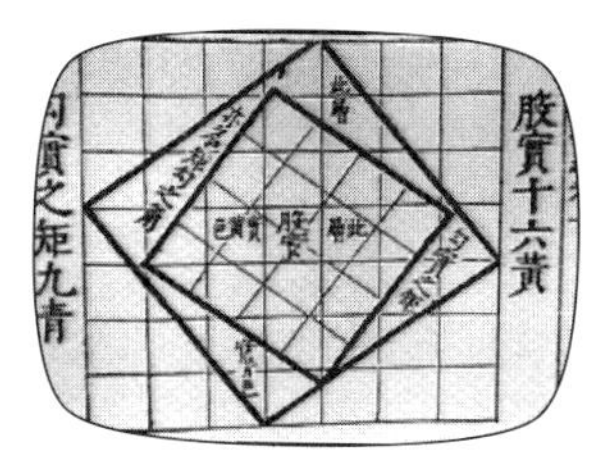

FIGURE 10. Chinese proof of the Pythagorean theorem.

orem. This is the proof usually given in textbooks and it is riddled with construction lines and many labels for points of intersection. The essence of Euclid's proof is greatly simplified through animation (Figure 11). Construction lines appear only when needed and then they disappear from the screen. By a combination of shearing and rotation, a blue square erected on one leg of the right triangle is transformed into a blue rectangle erected on part of the hypotenuse. The same process is repeated with a green square on the other leg which is transformed to a green rectangle erected on the rest of the hypotenuse. The blue and green rectangles together fill out a square on the hypotenuse, and the proof of the theorem is revealed visually.

The background color in the foregoing proof was selected to resemble a parchment page in an old book. This is the same color as the shearing demonstration in the review of prerequisites. To make the proof look more like Euclid's original proof, the shearing and rotation are applied to auxiliary triangles (half of the rectangles). We show the dark blue auxiliary triangle as rotating to two different positions; Euclid showed them in both positions overlapping each other. The animation avoids overlapping figures and is less confusing. When the triangle rotates, its edges are labelled with b and c to emphasize that the lengths of these edges do not change. Also there is a sound effect during the rotation that is different from the sound effect accompanying shearing.

A simplified variation of the proof is also shown that involves shearing without rotation. These proofs show the power of animation. The essence of the proof is revealed without cluttering the diagram with complicated construction lines and labels. The shear directions (parallel construction lines)

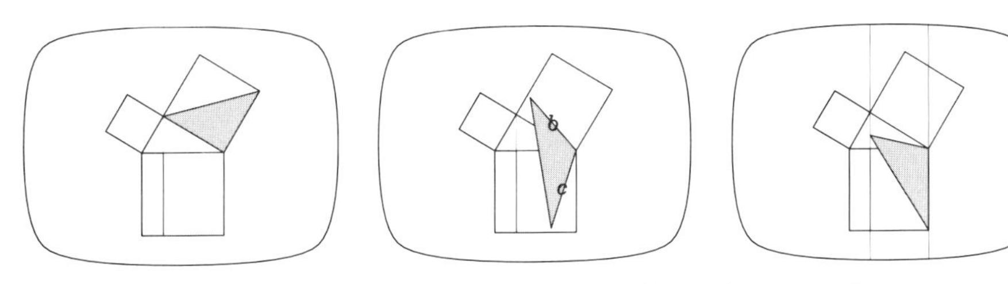

FIGURE 11. Animated version of Euclid's proof.

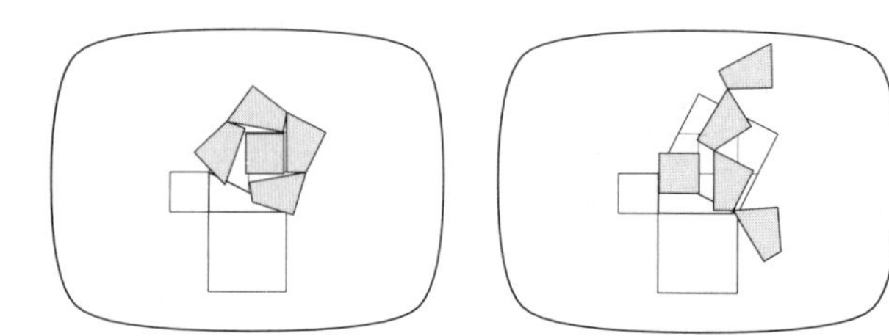
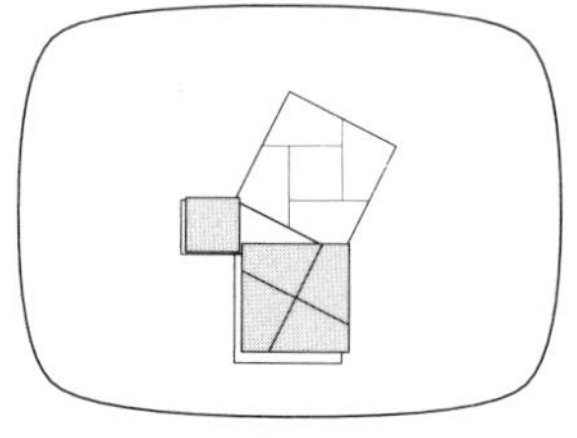

FIGURE 12. A dissection proof of the Pythagorean theorem.

are put on the screen as needed and then immediately removed.

We have found few professional mathematicians who can reconstruct the details of Euclid's original proof, but almost everyone, professional or amateur, understands and remembers the animated proof.

3.2.9 *A dissection proof.* The next segment shows a dissection proof in which the large square on the hypotenuse is divided into five pieces that are rearranged to fit exactly inside the other two squares (Figure 12). The division into pieces is done on the screen with a bright moving dot suggesting an electric spark or torch actually cutting out the pieces. An appropriate buzzing sound effect reinforces this image. As the five pieces are rearranged to form the smaller squares, a ghost image of their original positions remains on the larger square for reference. The pieces unfold as if they were connected by simulated hinge joints, and the corresponding sound effects (squeaky hinge joints) help to enliven the animation.

Audiences always laugh when they see this animated dissection proof. This particular piece of animation is not only entertaining but has pedagogical value. It gives teachers on opportunity to have students reproduce the dissection with scissors and paper. The workbook elaborates on this dissection proof and explains how infinitely many other dissections can be used to prove the Pythagorean theorem.

3.2.10 *Euclid's Proposition* 31. Next, the screen shows a diagram illustrating Proposition 31 in Book VI of Euclid's *Elements.* If the squares constructed on the edges of a right triangle are replaced by similar rectangles or any other three shapes similar to one another, the sum of the areas of the two smaller shapes is equal to the area of the larger shape. This is illustrated visually with various shapes (including cross sections of teapots) and an animated proof is given to explain why the theorem is true. Also, the argument is turned around to show that Proposition 31 is actually equivalent to the Pythagorean theorem. This is followed by a remarkably simple proof of the Pythagorean theorem based on Proposition 31. The original right triangle is divided into the same two similar right triangles that appeared in the algebraic derivation earlier in the program. Because the areas of the two smaller triangles obviously add up to the area of the larger one, the proof drops out immediately.

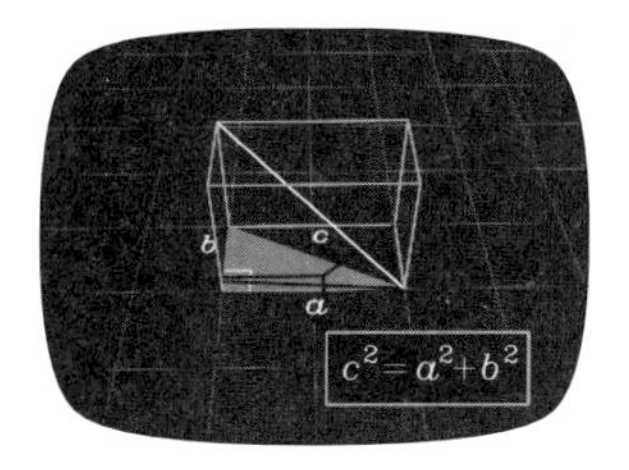

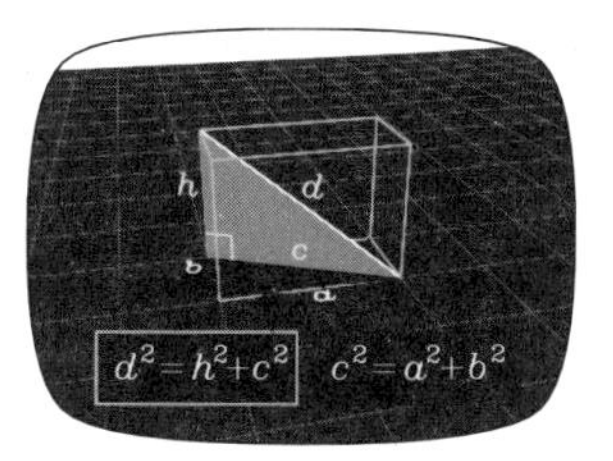

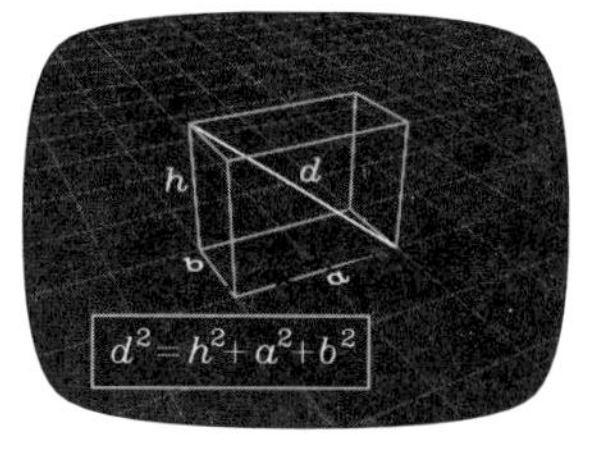

FIGURE 13. The three-dimensional version of the Pythagorean theorem.

There is potential for confusion in this scene because all three similar triangles used for the proof are inside the original triangle. Consequently they overlap and it is difficult to realize that three triangles are there. For this reason, a prologue shows pentagonal shapes flipping back and forth from outside to overlapping positions inside. This gets the viewer used to the idea that shapes can be inside the triangle. The shapes flip back and forth as though they were hinged along the edges of the triangle. Use of transparency helps keep overlapping figures distinct.

3.2.11 *Extension to three dimensions.* The power of animation is revealed again in the next segment, which extends the Pythagorean theorem to three-dimensional space. First, a rectangular building is shown. The roof and walls are removed, leaving only the skeletal outline of a rectangular box. The building and its skeleton cast a shadow on the ground plane to enhance the three-dimensional effect. The diagonal from one corner on the floor to the opposite corner on the ceiling is drawn and its length is calculated by two applications of the Pythagorean theorem. One right triangle appears on the floor, and another appears perpendicular to the floor. The two-dimensional Pythagorean theorem is applied to each of these and an algebraic ballet reveals the three-dimensional version (Figure 13).

Each line is drawn with darker edges so it looks like a thin cylinder. Thus, when one line crosses in front of another you can tell which is in front. All the labels in 3D cast appropriate shadows. This is difficult to choreograph so labels do not interfere with each other. Again, translucency of the triangles is used for clarity. Many teachers have commented on how difficult it is for them to display the essence of this proof on the chalkboard. They appreciate the fact that animation does it quickly and easily, in a way that all students can understand and remember.

3.2.12 *Previews of things to come.* The final segment of the videotape, a preview of things to come, contains an effective piece of three-dimensional animation intended to show that the Pythagorean theorem is a fundamental property of right triangles that lie in a plane. This is done by showing a right triangle lying on the surface of the earth. (Incidentally, the vertex of the right angle is located on the island of Samos, the birthplace of Pythagoras.) As the triangle gradually expands to cover a large portion of the earth's surface, it becomes a spherical triangle with three equal sides and three right angles lying on a dramatic rendering of the earth's surface. The Pythagorean theorem cannot hold for such a spherical triangle. The dramatic view of the earth serves as a background for the credit roll at the end of the videotape.

Throughout this and all videotapes in the series the visual images are enhanced by special sound effects and also by music, classical and modern, floating in the background. To attract students to an educational videotape, production quality must compete with commercial programs they see every day in their homes. Consequently, although the videotapes are designed to be used in the classroom, the production is of broadcast quality. This conveys a feeling of authenticity and importance that is difficult to achieve with videotapes of less than broadcast quality.

4. VIDEOTAPE ON *The Story of Pi*

4.1 Brief outline of the videotape on *The Story of Pi*. After a brief computer animated *Review of Prerequisites*, the program opens with a reporter interviewing several young people, asking, "What can you tell me about the number pi?" Each person gives a different answer, some of which are only partially correct. Later in the program the same people respond with correct answers.

The program defines pi as the ratio of the circumference to the diameter of a circle, then explains that pi appears in a variety of formulas, many of which have nothing to do with circles. After discussing the early history of pi, the program invokes similarity to explain why the ratio of circumference to diameter is the same for all circles, regardless of their size. This ratio, a fundamental constant of nature, is denoted by the Greek letter π. The formula $2\pi r$ for the circumference of a circle of radius r follows from the definition of π.

The program then turns to another ratio that is the same for all circles: the area of a circular disk divided by the square of its radius. By showing that a circular disk of radius r has area πr^2, Archimedes proved that this constant ratio is also equal to π.

Two animated proofs of the area formula are given. This is followed by the Archimedes method for estimating π by comparing the circumference of a circle with the perimeters of inscribed and circumscribed polygons.

The next segment describes a sequence of improved estimates for the value of π and points out that π is irrational. After demonstrating the presence of π in probability problems, the program returns briefly to the reporter who interviews the students again, asking, "*Now* what can you tell me about pi?" This time, each student gives a different correct statement about π. The concluding segment explains that major improvements in the estimates for π represent landmarks of important advances in the history of mathematics.

4.2 Comments on some of the visualization techniques used in *The Story of Pi*.

4.2.1 *What is the number pi?* When the interviews reveal that one cannot learn the meaning of pi by conducting a poll, the narrator explains that you can find out for yourself what pi is by measuring the circumference of a circle and dividing by its diameter. A student is shown on screen measuring two circular objects of different sizes. In each case, the measured circumference and diameter (approximate values) are displayed on the screen, and their ratio is calculated, giving two slightly different values, each greater than 3. The narrator explains that the ratio is always the same, no matter what size circle you use. This constant ratio is called π, the first letter in the Greek word for perimeter. The symbol π does a little dance accompanied by a squeaky sound effect that is introduced here and used repeatedly to call attention to π in later scenes. Inequalities on the screen show that π is somewhere between 3 and $3\frac{1}{7}$. Animation shows a ruler with a number line and an arrow labeled π pointing to a point between 3 and $3\frac{1}{7}$.

The foregoing segment on measurement was intended to stimulate discussion between teachers and students. The teacher, or an alert student, may well ask "How come she gets slightly different answers if the ratio of circumference to diameter is supposed to be the same?" This provides an opportunity for discussing the practical difficulties of making accurate measurements of real objects.

As the tape continues, the narrator explains that the program deals with two important questions. First question: How do we know that the ratio of circumference to diameter is the same for all circles, small or large? As the question is being asked, visual images are shown of circular objects in real life varying in size from small coins to a giant sun. Second question: How can we determine the exact numerical value of π? As this question is being asked, a dramatic animated zoom sequence shows the decimal expansion of π unfolding, digit by digit, as the camera zooms in on the ruler (Figure 14).

The motion of the digits of π was choreographed so the new digits keep appearing just over the arrow pointing to a fixed point on the number line. This allows students to focus on just that portion of the screen and see both

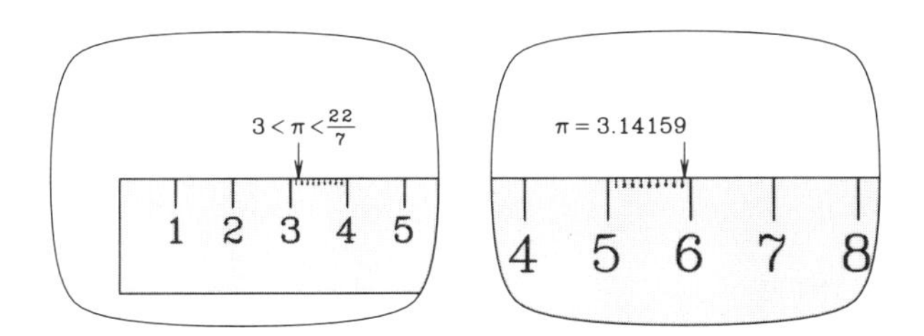

FIGURE 14. Animated zoom sequence revealing more and more digits in the decimal expansion of π.

the new digits and the portion of the number line from which they came. As the scene fades out, it leaves the impression that the digits will continue to unfold indefinitely. Many teachers have commented favorably on the effectiveness of this piece of animation.

The narrator continues by saying that we will also see why people have been interested in the number π from ancient times to the present day. The images here include shots of a Babylonian clay tablet and an Egyptian papyrus showing circular drawings related to early attempts to estimate the value of π, and also a news clipping from *Science News* describing a 1989 calculation of a half billion decimal places of π by the Chudnovsky brothers.

4.2.2 *Some uses of* π.

Areas and volumes of circular objects. The next segment, entitled *Some uses of* π, explains that π occurs in formulas for areas and volumes of circular objects. Animation shows the formulas for the area and circumference of a circular disk, and how the formulas change when the disk is transformed to a solid cylinder. The same is done for a circular cone and a torus. This was originally planned to be a series of equations containing the number π, but it soon grew into a mini-geometry course on areas and volumes. As the two-dimensional shapes move to become three-dimensional objects, the equations for the surface areas or volumes of solids so generated change according to Pappus' rules. For example, a circular disk is shown with circumference $2\pi r$ and area πr^2. The disk is moved a distance h to sweep out a solid cylinder whose volume is $\pi r^2 h$ and whose lateral surface area is $2\pi rh$. The circle itself is shown in red and its interior in yellow. When the cylinder is formed, it has the appearance of a cylindrical block of yellow cheese wrapped in a red skin. Similarly, when a semicircular disk is rotated to sweep out a solid sphere, the formula for the volume is obtained by multiplying the area of the semicircular disk by the length of the path of its centroid. As the shape sweeps around from 0 to 2π radians, the corresponding term in the formula grows from a flat line to the expression " 2π ". The colors, yellow interior and red exterior, are meant to look like an apple with red skin. Because many things are being shown on the screen at the same time, we call attention to the symbol π by having it perform its squeaky dance.

Orbits. A shot of the earth shows the great circle route of an airliner flying from New York to Tokyo, as the narrator explains that π appears in the formula for the flight time. The animation was timed for the approximate duration of flight relative to the rotation of the earth. The movement of the sun as a light source is correct, but to make the trajectory show up better, the night side is artificially brighter than the actual night side would be.

Another piece of animation shows planets in their nearly circular (elliptical) orbits about the sun, with the dancing number π appearing in the formula for the time it takes a planet to make an orbit around the sun.

Gaussian curves. Next it is shown that π occurs in situations that have nothing to do with circles. Balls are shown falling at random in an apparatus to form a pattern much like a Gaussian curve, and animation shows that the area of the region under a particular Gaussian curve is equal to the square root of π. This is achieved visually by allowing the colored region under the curve to pour like a liquid into a rectangular region whose width is 1 and whose height equals the area $\sqrt{\pi}$.

4.2.3 *History of* π. This historical segment describes early attempts to estimate the value of π, including the Babylonian estimate 3.125, the Egyptian estimate 3.16, and the Biblical estimate 3. Animation on the number line shows the relative positions of these estimates. The original idea was to zoom in on better and better approximations as history progressed, but unfortunately the improvements were not a monotonic function of historical time. Since we were not showing a continuous zoom into the number line, extra labels in little boxes were inserted along the line to indicate the values at the corresponding tick marks.

This is followed by an animated segment explaining that similarity shows that the ratio of circumference to diameter is the same for all circles. This ratio is a fundamental constant of nature that we call π.

4.2.4 *A discovery of Archimedes.*

Area of a circular disk determined by dissection into radial slices. The next segment explains that similarity also shows that the ratio of the area of a circular disk to the square of its radius is another fundamental constant, which, as Archimedes showed, is also equal to π. To explain why, the area of a circular disk is calculated by dividing the disk into a large number of radial slices. The slices are rearranged to form a new figure that is almost a parallelogram with the same area as the disk. As the number of slices increases, the graphics convince the viewer that the parallelogram becomes more and more like a rectangle with base πr and altitude r, so the area of the circular disk is πr^2 (Figure 15).

Area of a circular disk determined by dissection into concentric rings. The same result is obtained by another method. This time the disk is divided into equally spaced concentric rings. The outer ring is unwrapped to form a rectangular strip whose base is nearly the same length as the circumference and whose area is equal to that of the outer ring. The same is done with all the rings, and they are stacked into a pile that looks somewhat like a right triangle

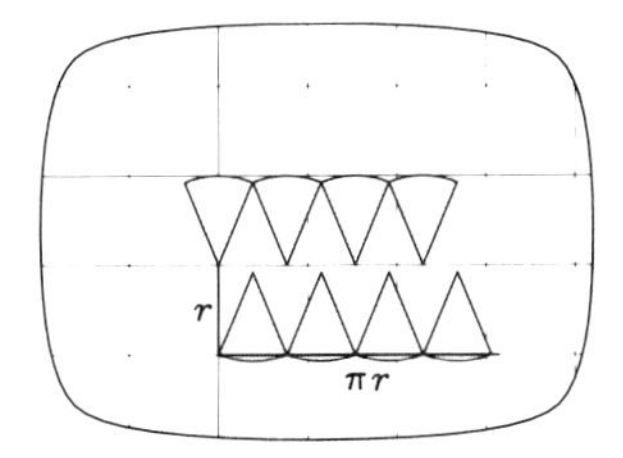

FIGURE 15. Animated derivation of the formula for the area of a circular disk.

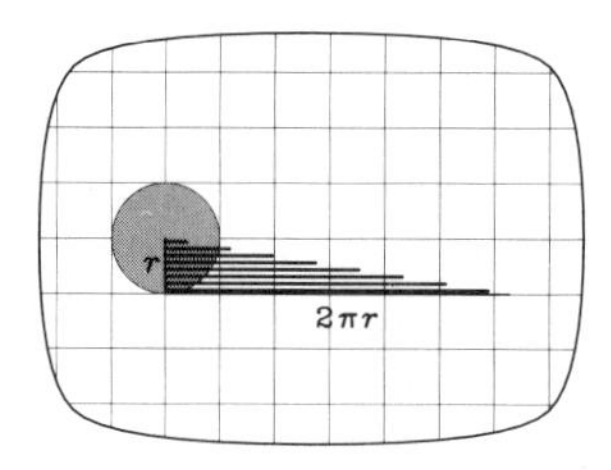

FIGURE 16. Another method for calculating the area of a circular disk.

(Figure 16). The base of the triangle is nearly the circumference of the circle, and its altitude is equal to the radius. As the number of rings increases, the stack of rectangular strips looks more and more like a right triangle of base $2\pi r$ and altitude r, so once more we see why the area of the circular disk is πr^2. We regard these last two animated segments as conclusive proof that sophisticated ideas of calculus can be introduced at an early stage in a child's education through the imaginative use of computer animation. When the student is introduced to these concepts again in the formal context of a calculus course, the images in these segments will demonstrate visually and dramatically the idea of a limit process.

4.2.5 *Computation of* π.

Archimedes' method. The program then describes Archimedes' method for estimating π by comparing the circumference of a unit circle with the perimeters of inscribed and circumscribed polygons. Archimedes started with regular hexagons and obtained the inequalities $3 < \pi < 2\sqrt{3}$. This is shown visually by unwrapping half the circumference of a circle and the corresponding portions of inscribed and circumscribed hexagons. Animation shows how Archimedes kept doubling the number of sides until he reached a regular polygon of 96 sides and obtained the rational estimates

$$3\tfrac{10}{71} < \pi < 3\tfrac{1}{7},$$

giving π correct to two decimals.

Later estimates. Diagrams from old Chinese manuscripts are shown as the narrator explains that many centuries later the Chinese found the rational approximation $\frac{355}{113}$, which gives six decimals of π. This was the world's record for more than a thousand years until the use of Arabic numerals provided more efficient ways of doing arithmetic. The narrator explains that when infinite series and trigonometric functions became widely known, formulas were discovered that made it possible to approximate π without geometric diagrams. Some of these formulas are shown on the screen. By the end of the nineteenth century, these formulas were used to calculate π to hundreds of decimal places. In the twentieth century, high-speed electronic computers were coupled with new mathematical methods that enabled computers to do arithmetic with very long numbers. The screen reveals an article from *Science*

News describing a calculation by the Chudnovsky brothers. The narrator then asks, "Why go to the trouble and expense of hooking up giant supercomputers to calculate a billion decimals of π?" As the camera pans over several supercomputers and their internal hardware, the narrator explains that these calculations are used to test the architecture of supercomputers. Computing a million digits of π gives large computers a thorough workout and provides a good measure of the computer's overall efficiency. The calculation also checks software and programs for speed and accuracy.

Irrationality of π. We can also learn more about π itself. The decimal version of $\frac{22}{7}$ appears on the screen as the narrator explains that for a long time people wondered if π was an exact fraction (a rational number) like $\frac{22}{7}$. It is pointed out that rational numbers always have repeating patterns in their decimal expansion. As people calculated more and more decimals for π they searched for repeating patterns, but none were ever found. Many decimals of π are shown on the screen with a scanner searching unsuccessfully for repeating patterns. A portrait of Johann Lambert appears as the narrator explains that in the eighteenth century Lambert proved that π is irrational, so no repeating patterns exist. As the number line appears, once more showing the relation of π to the Archimedes estimates, the narrator explains that rational approximations are quite useful. The Archimedes estimates are closer to π than any other fraction with a denominator less than 100, while the Chinese estimate $\frac{355}{113}$ is closer to π than any other fraction with a denominator less than 16,000.

4.2.6 *Further uses of* π.

Visibility of lattice points. The next two scenes show two examples of probability problems in which the number π occurs unexpectedly. The first example describes visibility of lattice points from the origin. An effective use of computer animation reveals that about $\frac{2}{3}$ of the lattice points in the plane are visible from the origin, with the exact probability being $\frac{6}{\pi^2}$ that a lattice point chosen at random will be visible from the origin (Figure 17).

Buffon needle problem. The second example is a special case of the Buffon needle problem, effectively illustrated with animation. A large number of needles of the same length are shown falling at random on a pattern of

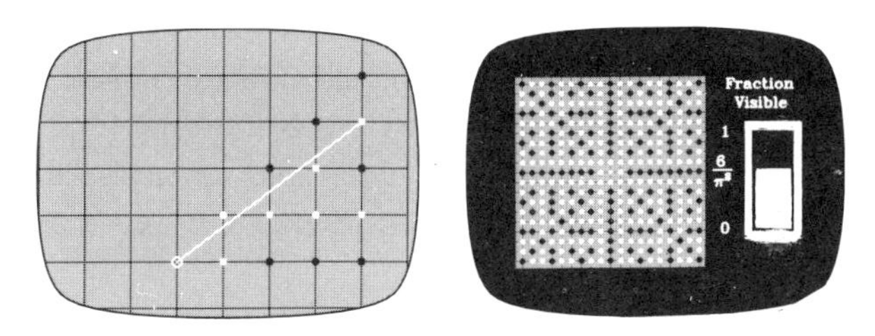

FIGURE 17. Animation showing lattice points visible from the origin.

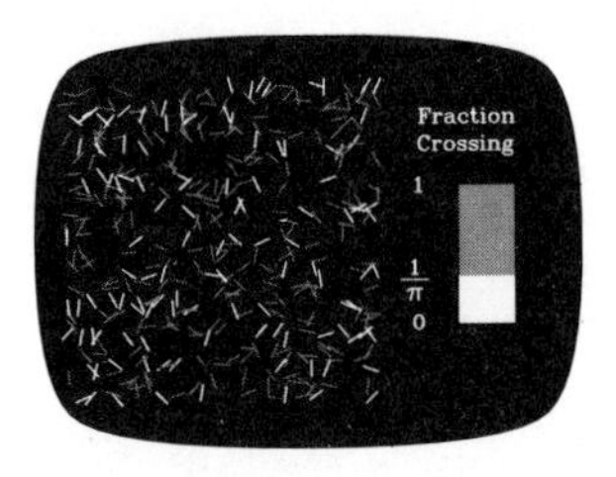

FIGURE 18. The Buffon needle problem.

equally spaced parallel grid lines whose spacing is twice the length of the needles. About one third of the needles cross a grid line (Figure 18). The exact probability that a needle will cross a grid line is $\frac{1}{\pi}$.

4.2.7 *Recap.* In the final segment, the students are asked again to tell what they known about π, and this time the answers suggest that they have learned something from watching the program. A brief recap of the program points out that advances in the computation of π have been markers of significant progress in the history of mathematics. People continue to calculate more and more digits of π because it is a challenge to the human spirit, like climbing Mt. Everest or traveling to the outer planets. A dramatic shot of planet Neptune serves as a background as the narrator concludes with the statement that π, like the outer planets, is built into the fabric of our physical universe and will always be explored.

5. VIDEOTAPE ON *Polynomials*

5.1 Brief outline of the videotape on *Polynomials*. This program provides a visual catalog of the shapes of graphs of polynomials in a rectangular coordinate system. The program opens with examples of polynomial curves that appear in real life, including parabolic orbits of projectiles in sports and cubic splines used in designing sailboats.

This is followed by a systematic description of polynomials by degree. Linear polynomials are discussed first; their graphs are straight lines of various slope. Quadratic polynomials are discussed next; the prototype being the parabola with Cartesian equation $y = x^2$. It is shown how the Cartesian equation changes if the curve is translated vertically or horizontally and also how the equation is altered by a vertical change of scale. The general quadratic polynomonial is obtained from the prototype $y = x^2$ by a combination of horizontal and vertical translation and vertical scaling. The next section deals with intersections of lines and parabolas and leads naturally to zeros of polynomials.

Cubic polynomials are treated next, with discussion of possible zeros, local maxima, local minima, and points of inflection. Cubics have three prototypes, $y = x^3$, $y = x^3 + x$, and $y = x^3 - x$, from which all others can be obtained by horizontal or vertical translation, by horizontal or vertical change of scale, or by taking mirror images. A similar discussion is given for quartics and higher degree polynomials, all of which have infinitely many prototypes.

Many curves that are not graphs of polynomials can be approximated with great accuracy by polynomial graphs. An engaging computer animated segment shows how polynomial approximations to a sine curve give greater accuracy as the degree increases.

5.2 Comments on some of the visualization techniques used in *Polynomials*.

5.2.1 *Linear polynomials.* The main purpose of this section is to introduce the concept of the slope of a line, change in height divided by change in horizontal distance, and to show how slope enters the linear Cartesian equation $y = mx + b$ as the coefficient multiplying x. An effective device that displays the visual meaning of the coefficients m and b is an animated hand turning a crank that changes each coefficient independently. For fixed slope m, the crank changes the constant term and the line moves up or down parallel to itself. Then the constant term b is kept fixed, and the crank changes the coefficient of x to show how the slope changes, first through increasing positive values, then through decreasing positive values to zero, and then to negative values (Figure 19). The crank is used throughout this program.

5.2.2 *Quadratic polynomials.* This segment starts with the graph of the quadratic polynomial equation $y = x^2$. A constant term is appended, and as a hand crank changes this constant the curve is translated up or down. Then we see what happens to the equation if the curve is translated horizontally. The new equation has the form $y = (x - h)^2$.

Next, a change of scale in the vertical direction changes the shape of the parabola. The parabola opens upward if the highest power x^2 is multiplied by a positive factor, and it opens downward if x^2 is multiplied by a negative factor. Again, the hand crank is used to show the changing coefficients. Finally, all three operations are performed, horizontal translation by h, vertical scaling by a factor a, and vertical translation by k. An algebraic ballet shows that the resulting equation $y = a(x - h)^2 + k$ always has the form $y = ax^2 + bx + c$, a quadratic polynomial in x. Again the hand crank is

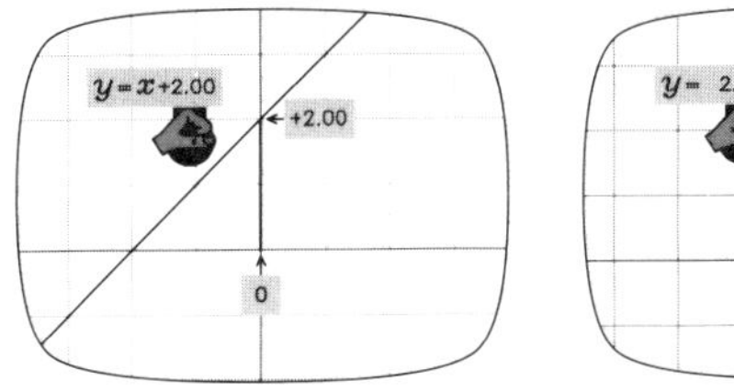

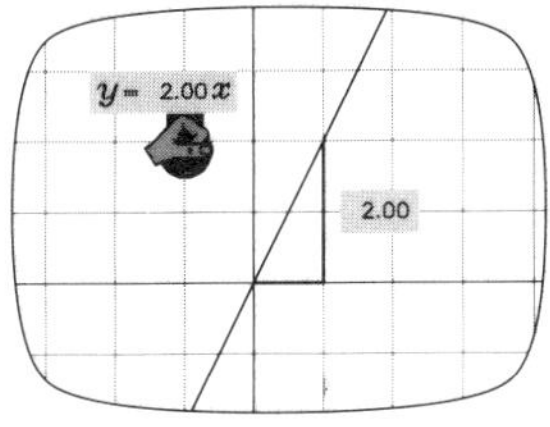

FIGURE 19. Showing how the graph of $y = mx + b$ changes when the coefficients b and m change.

used to show what happens to the graph if you change the constant term, the coefficient of x, and the coefficient of x^2.

In this and several other scenes, a distinction was made between labeling particular points on a graph and labeling the function represented by the graph as a whole. Individual x or y coordinates were labeled with green slates, the color of the background grid. The entire graph of the function was labeled with a gray slate, the color of the line. This helps emphasize to students the difference between a function, represented by the entire graph, and the value of the function at particular discrete points.

5.2.3 *Intersections of lines and parabolas.* This segment begins by finding the points of intersection of a line $y = mx$ with the parabola $y = x^2$. An algebraic ballet reduces this to the problem of solving the quadratic equation $mx = x^2$, whose roots are $x = 0$ and $x = m$. Examples are given to show that a line may or may not intersect a given parabola, and that the number of intersection points can be zero, one, or two. When the x-axis intersects the parabola, the points of intersection are called real zeros of the quadratic. Examples are shown with two distinct real zeros, two equal real zeros, and no real zeros. Sound effects accentuate the appearance and disappearance of zeros. An application describes how these ideas are used in constructing computer animated images of planet Jupiter.

5.2.4 *Cubic polynomials.* Three prototypes, $y = x^3 - x$, $y = x^3$, and $y = x^3 + x$, reveal all the basic properties of cubic curves. Each graph has two parts, one that bends upward and one that bends downward. The point where the upward and downward parts meet is called a point of inflection. Again, the hand crank is used to change the coefficients. Changing the constant term translates the graphs up or down and reveals that the number of real zeros can be 3, 2, or 1. Any cubic polynomial can be changed to one of these three prototypes, or a mirror image, by changing its coefficients.

Next, by relating zeros with factors, it is shown visually why every cubic has at least one real zero. All cubics with three distinct real zeros have essentially the same basic shape, with a local maximum and a local minimum lying between consecutive zeros. Animation shows how the shape changes when the coefficients are adjusted to bring two of the zeros into coincidence. And when all three zeros are brought into coincidence, the local maximum and minimum coalesce to form a point of inflection. In this and several other scenes, the polynomial is displayed both as a linear combination of powers of x and also in factored form. The hand crank is used here to show adjustments in the coefficients of the factors. At the same time, the coefficients of the powers of x are continuously calculated and displayed.

5.2.5 *Polynomials of higher degree.* A corresponding discussion takes place for quartic curves, graphs of polynomials of degree 4. There are three symmetric prototypes,

$$y = x^4 - x^2, \quad y = x^4, \quad \text{and} \quad y = x^4 + x^2,$$

together with an infinite family of asymmetric prototypes

$$y = x^4 + rx^2 + x,$$

where r, the coefficient of x^2, ranges through all real numbers. Again, the hand crank illustrates that every quartic curve can be converted to one of these prototypes, or a mirror image, by adjusting the coefficients. Polynomials of degree 4 can always be factored into two quadratic factors, and the total number of zeros, determined by the zeros of the quadratic factors, can be 0, 1, 2, 3, or 4.

As the degree increases, the number of prototypes also increases. The maximum number of real zeros is equal to the degree, and the number of peaks and valleys is at most one less than the degree. Polynomials of odd degree always have at least one real zero.

9. Concluding Remarks

Words and pictures are different mechanisms with a single purpose—the presentation of information. Visualization—the visual representation of ideas, principles, or problems—has always played an important role in both teaching and learning mathematics. Visualization is even more effective when the images are in motion. This article explains how *Project MATHEMATICS!* combines words and moving images, together with music and special effects, to produce computer animated videotapes on basic topics in high school mathematics.

Videotapes provide a highly efficient way of conveying information. If one picture is worth ten thousand words, a 20-minute videotape must be worth millions of words. However, the real value of video technology is not the efficiency by which it transmits information, but the manner in which the information is transmitted. Visual images make a much greater impact than printed or spoken words. People tend to forget words they hear or read, but images are retained for a long time because they have emotional as well as intellectual appeal. This is especially true of moving images accompanied by music and sound effects. These videotapes show that the full power of modern video technology can be put to work to provide a valuable pedagogical tool that reveals mathematics for what it is, not only understandable and exciting, but eminently worthwhile as well.

Because video transmits a large amount of information in a relatively short time, it is not expected that all students will understand and absorb all the information in one viewing. The programs are intended to be seen repeatedly. Video technology makes it possible to pause and back up to review a scene. Also, tapes can be borrowed from a school's video library for private viewing. Students who lack language skills, or who are less apt to respond quickly in class, gain confidence in their abilities by viewing videotapes privately. *Project MATHEMATICS!* allows its videotapes to be copied without charge when they are used for educational purposes, so teachers and students can make their own copies for personal use at a low cost.

NOTE. Tapes and workbooks can be obtained at nomimal cost from the Caltech Bookstore, 1–51 Caltech, Pasadena, CA 91125; telephone (818) 395-6161.* Further information about the project can be obtained by writing to the authors at *Project MATHEMATICS!*, 1–70 Caltech, Pasadena, CA 91125.

PROJECT MATHEMATICS!, CALIFORNIA INSTITUTE OF TECHNOLOGY, PASADENA, CA 91125
E-mail address: apostol@caltech.edu
E-mail address: blinn@caltech.edu

* Until July 1, 1993, please call 818-356-6161.

CBMS Issues in Mathematics Education
Volume 3, 1993

Calculus for High School Teachers: Content, History, Pedagogy

PETER BRAUNFELD

More and more high schools are offering calculus courses to their students. Such courses come in a wide assortment, ranging all the way from demanding courses in AP calculus to "quickie" introductions to the basic ideas or, regrettably and all too often, just the basic techniques without the ideas. There has been and continues to be some debate about the desirability of teaching calculus in high school. But I think that this debate, mainly among college professors, is largely ex post facto, and I see no reason to believe that the current trend is likely to be reversed.

Given that high school calculus is here to stay, it becomes important to make sure that high school teachers are properly prepared to teach it. One of the problems is that so many high school teachers have not thought about or done calculus for a very long time. Their calculus is rusty and they are, quite rightly, afraid to launch into teaching it.

In view of all this, I decided in 1989 to develop an in-service calculus course for teachers. As it turned out, this course has also attracted many teachers of precalculus courses, who report that studying the ideas of calculus really helps them put in focus the mathematics they are currently teaching.

The big question, of course, is: What should a calculus course for teachers look like? Obviously, it should not just be a rehash of the typical introductory freshman calculus course. The primary reason is simply that teachers need to know much more than what they will be teaching to others. Secondly, as a practical matter, if an in-service course is to attract students, it needs to carry graduate credit!

There is a long-standing, traditional answer to the question: What do you need to know to teach calculus properly? Answer: a beginning course in real variables. The argument goes something like this: to "really" understand how and why calculus works, we need, first, to clarify and make precise the basic ideas, like limit, derivative, integral, etc. On this basis, we should then derive, in a detailed and logical fashion, all the standard theorems. This will

©1993 American Mathematical Society
1047-398X/93 $1.00 + $.25 per page

allow the student to see how the subject "hangs together", i.e., what follows from what, what is needed to prove what, and so forth. Indeed, I have heard it argued that formulating precise definitions and exposing the logical structure underlying *any* piece of mathematics is precisely what we *mean* by "understanding".

There is certainly some merit in the notion that someone who has worked through the "epsilontics" of limits or seen in some detail, say, the role of continuity in the development of calculus, has gained some very valuable insights into its workings. In practice, however, my experience over many years has been that this neat theory simply does not work. Not only is the typical junior/senior real variables course dreaded by most pre-service mathematics teachers, but, so often, they seem to emerge from the encounter very little wiser than before coming in. So many of them simply never seem to get the point of what it is we are trying to get them to understand or that this "what" even needs understanding. All too often, they get so mired in the details that they never see the "big" picture. As they slog through trying to find the "delta that goes with a given epsilon", they cannot remember why they are trying to find this delta or what it tells them when they do find it. Worse yet, in my experience many students never really come to appreciate how it is that all these epsilons and deltas put the calculus they have learned on a "firm" foundation; in fact, they do not seem to feel that the calculus is much in need of a firmer foundation! To be sure, all this is most regrettable, but I really believe it describes the situation fairly.

Sometimes, as at my university, we try to enhance the preservice teachers' understanding and appreciation of calculus by also requiring, or strongly recommending, a course in the *history* of calculus. Such a course certainly does provide useful background, and, by and large, students do better with it than the real variables course. However, I believe that a good history course really presupposes that the students have their calculus still very fresh in their minds. I also feel that teachers about to teach their own calculus course need more than just historical perspective (although my course does include quite a number of "historical" topics).

The course I finally developed covered, by and large, the topics of a typical first semester calculus course but from a more advanced standpoint. It was based on six strands or themes:

1. Content analysis. We examine in some detail the basic ideas of calculus (function, limit, continuity, derivative, rate of change of a function, etc.). What is important here, in my view, is not so much formulating precise definitions (using epsilons), but rather developing a rich and sound intuition about these ideas. The kind of content analysis I try to do, though not so popular in the United States, has a long tradition in Germany, where it is called "Didaktik der Mathematik" and goes back at least to the work of Felix Klein [**1**]. In a didactic analysis you look at mathematical ideas (e.g.

function) from a variety of points of view: intuitive, graphic, application, history, formal, etc., especially with the aim of acquiring a rich repertory of teaching strategies. Among the didactical questions the teachers and I discuss in my course are: What ideas and techniques in calculus should be stressed and which can be given shorter shrift? What are the pros and cons of various ways to sequence topics? What should be the role of proofs? When do proofs help and when are they best suppressed?

2. Historical perspectives. Often, to understand a mathematical idea, it is very helpful to trace its development over time. Nevertheless, as I mentioned above, I did not want just a history of calculus course. Instead I followed the example of Toeplitz [2], who coined the very apt word "genetic" to characterize an approach where one may pick and choose, even adapt, one's history to the purpose at hand. In the genetic approach one has, for example, no problem skipping over large chunks of history or recasting cumbersome ancient arguments in modern notation.

3. Applications. I collected from many sources the most interesting and challenging problems I could find. Mostly, I had the teachers work on these problems in collaborative learning groups of threes and fours (see below).

4. Pedagogical analysis. Here we tried to look at calculus from the student's point of view, e.g., what is easy and what is hard? In the course of teaching calculus on and off over some thirty years, I have collected some interesting data about such questions—data on the kinds of mistakes students make and the kinds of misconceptions they have. We discussed some of these often surprising statistics and tried to understand how they might inform our teaching strategies.

5. Applications of technology. As in every branch of mathematics, the advent of calculators and computers has, or should have, a dramatic effect on our teaching of calculus. In this course we focused mostly on classroom uses of graphing calculators.

6. Applications to the high school classroom. Since my purpose was to have an impact on high school classrooms, I wanted the teachers to work with some aspect of what they were learning in that context. To this end, I required each teacher to develop some kind of lesson or class activity based on course material, and use it with kids. These projects had to be documented in writing, as well as presented orally to the class.

The course has been taught in two very different formats: (1) as an intensive residential two-week summer institute (six hours of daily instruction plus homework), followed by a weekend workshop in the spring in which teachers presented their classroom projects; and (2) as an on-site course taught one evening (three hours) per week for one semester. (In this format, the teachers did their projects during the semester itself and presented their reports at the last sessions.) There are advantages and disadvantages to both these formats, but it turns out that one can cover just about the same amount of material in either of them.

In both formats, the structure of each lesson (a three-hour chunk) is more or less the same:

1. Introduction of new material: This is what might be construed as the lecture component of the course, but it was really more of a mix of "guided discovery", Socratic dialogue, discussion, and class exploration of the material. The fact that the classes were small (circa fifteen) made this easy!

2. Problem sessions: Students were given problem sets and asked to solve them during class time in small groups. I stressed the importance of making these activities a genuine *team* effort. Problems not completed by the teams during the class period were left for individual homework.

3. Wrap-up: The class came together once more as a whole to discuss the events of the day. Teams exchanged information on their efforts to solve the problems; in addition, we reviewed the material and tried to resolve outstanding student questions.

Below is a course outline—annotated at several places with examples of how some of these topics were actually handled and illustrating each of the five themes referred to above.

Course Outline

1. Functions. I point out the value of thinking of a function as a machine. A slight variation from the standard function machine is that mine has not only the usual input and output slots, but a little red light on the front (Figure 1). The machine accepts any input whatsoever. For a given input x, one of two things happens: either $f(x)$ comes out the output slot, or the red light goes on (indicating that this is an "illegitimate" input, i.e., one the machine cannot properly process). The red light leads very naturally to the

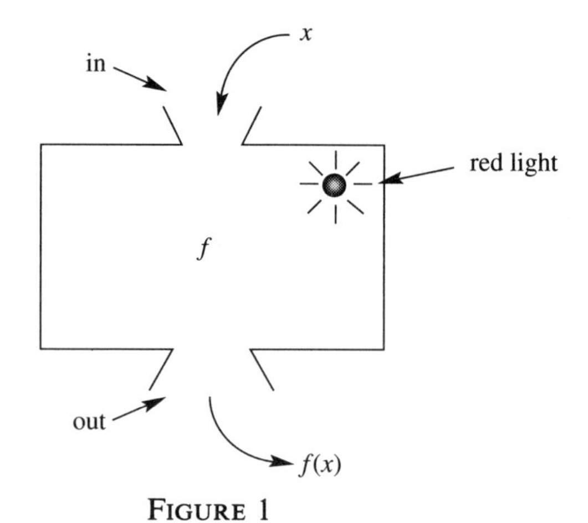

Figure 1

notion of the *domain* of a function (i.e., the set of inputs that do not activate the red light). The notions of range, piece-wise defined functions, restricted domains, composition of functions, and inverse functions all have natural interpretations in the machine model.

2. Limits. The most important idea here is $\lim_{x\to a} f(x)$ (for derivatives, we do not need limits of sequences, only the idea of the limit of a function at a point a). I emphasize that we are dealing here with two basic, mutually complementary ideas which should be held up for comparison and contrast:

$$f(a) \quad \text{and} \quad \lim_{x\to a} f(x).$$

$f(a)$ tells us about the behavior of the function $f(x)$ *at* a; it tells us nothing about the behavior of $f(x)$ *near* a.

$\lim_{x\to a} f(x)$ tells us about the behavior of the function *near* a; it tells us nothing about the behavior of $f(x)$ *at* a.

To illustrate these ideas, let us imagine the following game to be played with a computer. The user is told that the computer is thinking of some function $f(x)$ and the user's job is to guess the value of that function at, say, 2. The user may ask the machine to evaluate the function at any point *other* than 2. It would seem the following is a reasonable strategy to pursue: ask the machine for values very close to, but different from, 2. If these values do not seem to converge then it is hopeless to predict the value of f at 2. On the other hand, if there does appear to be a limit, that still does not guarantee that this limit and the function-value at 2 will be the same. But it does seem that at least this limit is a reasonable prediction for that value and, indeed, if there are no surprises at 2, we say the function is *continuous* at 2.

Now we ask: How can it happen that $\lim_{x\to a} f(x)$ fails to exist? Most commonly, if the left- and right-hand limits at a exist, but fail to agree (a jump discontinuity). Cases where even the one-sided limits themselves fail to exist are fairly rare, but they do happen and we take note of them (e.g., $\sin(1/x)$ at 0).

3. Tangent problems. The derivative is often introduced as the way to solve tangent problems. I point out, however, that many tangent problems can be easily solved without derivatives, and that before the advent of calculus, tangents were often found using just algebra. The example in Figure 2 (see next page) shows how to find the tangent to $y = x^2$ at any point—without calculus. (It generalizes easily to: $y = x^n$). We consider the pencil of secant lines through P. For the tangent, we need $P = Q$, i.e., we need a double root at $x = a$. We have: $x^2 - a^2 = m(x-a)$ or $(x-a)(x+a) = m(x-a)$. We already know a is a root, so cancel $x - a$, to get $x + a = m$. Now find m so that a is a root again: $a + a = 2a = m$, so the solution is $m = 2a$.

4. Definition of derivative. I introduce, inter alia, some alternative definitions of the derivative, e.g., $f'(x) = \lim_{h\to 0}[(f(x+h) - f(x-h))/2h]$. We discuss

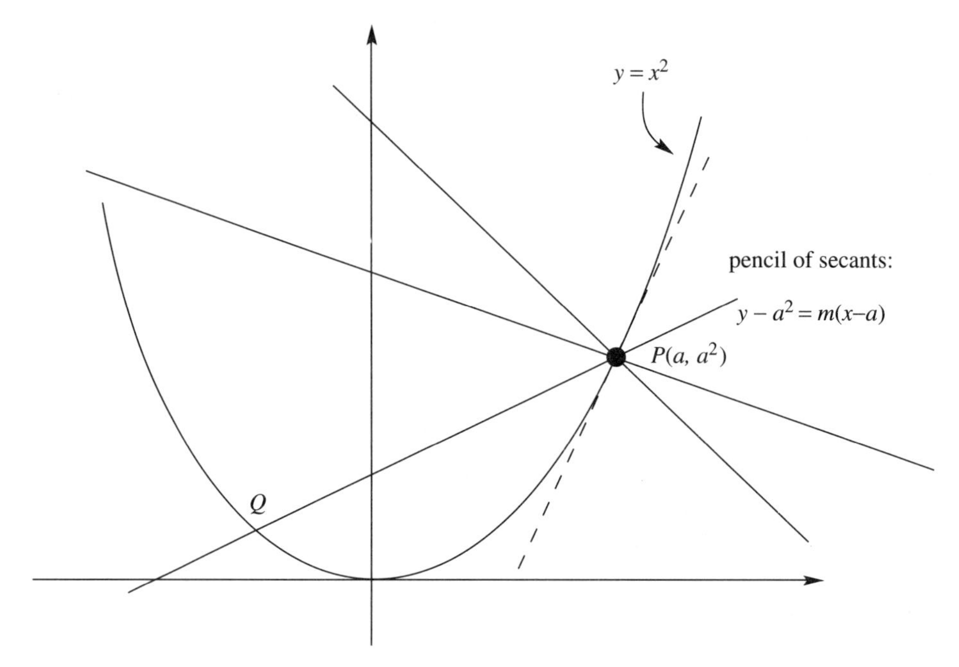

FIGURE 2. Pencil of lines through P

the pros and cons of such alternative definitions and whether or not they are equivalent to the standard definition. We also examine with some care the relation of differentiability to continuity at a point, using pictures, rather than proofs, to understand what is going on.

5. Differentiation rules. We look at the standard rules for sums, products, etc., as well as the chain rule, implicit differentiation, and inverse functions. Pictures are stressed wherever possible.

6. Related rate problems. Even the old war horses can be looked at from fresh points of view. Consider for example the following classical ladder problem.

A ladder of length p leans against a vertical wall as shown in Figure 3. If the bottom of the ladder is sliding away from the wall at a rate of a ft/sec., how fast is the top of the ladder sliding down the wall at any given moment?

The solution is

$$\frac{dy}{dt} = -\frac{x}{y}\frac{dx}{dt} = -\frac{x}{y}a.$$

At this point, traditionally you would hand in the problem for grading and go on to the next. However, if you pause to reflect on the solution, you might notice that $\lim_{y\to 0} dy/dt = \infty$. How can this be? Surprisingly, many students feel, at least at first, that this might actually be a physically realizable possibility! After we become convinced that this really cannot happen, we discuss what that tells us about the solution, and in fact, about the problem itself. Are the problem and/or solution completely meaningless? Can they be "rescued"?

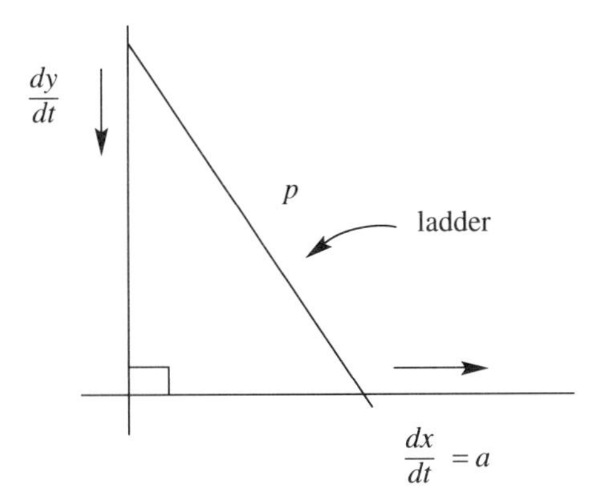

FIGURE 3

7. Max/min problems. New issues arise here from new technology; using spreadsheets and other software packages, we can now solve such problems with very good approximations and with no calculus. What is the role of calculus vis-a-vis such problems in the light of the new technology?

8. Mean-Value Theorem. There are, I think, three important things to discuss in connection with the Mean-Value Theorem.

1. Why is it important? In older texts, this theorem was sometimes called the Fundamental Theorem of Differential Calculus. Why do we make such a fuss about this theorem? Where is it needed and how is it used?

2. Interesting interpretations: What does the theorem say about speedometers on cars, average speeds, etc.?

3. Applications: the Mean-Value Theorem gives neat, quick proofs of a number of nice inequalities, e.g., $|\sin a - \sin b| < |a - b|$.

9. Curve-sketching. The hand-held graphing calculator will surely have a profound effect on the role of curve-sketching, just as the four-function calculator earlier affected the teaching of the arithmetic algorithms in elementary school. It seems to me that the emphasis now shifts from curve-*sketching* to curve-*reading*. In the former case, we use the analytical expression for the function to calculate its derivatives, and then use these to draw a picture of the function. In the latter case, we use the picture of the function to draw conclusions about the behavior of the function and its various derivatives. I believe that curve-reading in this sense is a very important and undertaught skill, and something the machines cannot (yet) do for us.

A bit of personal history: in the old days, long before calculators, I always gave a curve-sketching problem on the final exam. I dreaded grading these problems for a number of reasons: (1) if the student made any kind of mistake toward the beginning of the calculation, say, in calculating the first derivative, then, of course, everything from that point on would also be incorrect and, what's worse, inconsistent. To assign any kind of partial

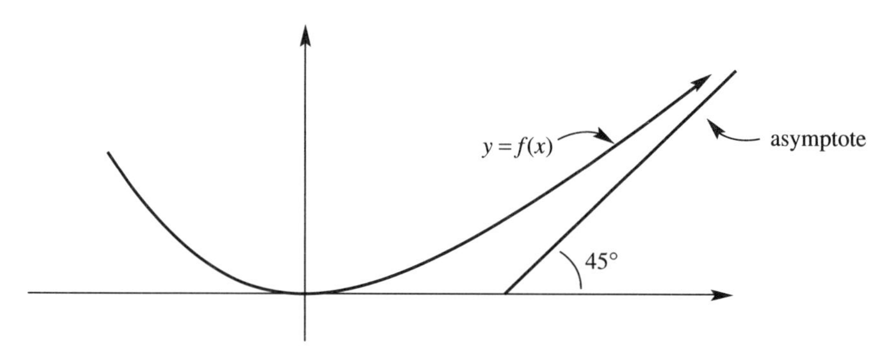

FIGURE 4. Question: What is $\lim_{x\to\infty} \frac{dy}{dx}$?

credit, one was faced with an arduous, unpleasant task of untangling a horrible rat's nest. (2) Even students getting the calculations right all too often drew bizarre graphs. I could never quite decide whether this was due to neurological problems with small muscle control in their hands or conceptual misunderstandings. As classes in the 1960s grew steadily bigger, I decided one day, just as a matter of self-preservation, to invert the graph-sketching question and turn it into a set of graph-reading questions instead. I fretted that the new questions might be much too easy, but to my great surprise, students had, and continue to have, great difficulties with this type of question. I give one example in Figure 4. One out of three students missed this question on an exam in 1972! Among the more popular incorrect answers were: 0, ∞, 45°, $\frac{\pi}{4}$.

10. Area. Just as tangent problems are the traditional entree into the derivative, so area problems are used to introduce the definite integral. Here, I believe, the historical, or better, the genetic, approach is particularly helpful. I begin with the Greeks, pointing out that they often thought of area problems as construction problems, i.e., construct a square equal in area to that of a given figure. (The best known problem of this type is, of course, the famous problem of squaring the circle.)

I begin with an elementary problem that uses only high school geometry: Find a square equal in area to a given rectangle. If the rectangle has length a and width b, then the desired square has side $s = \sqrt{ab}$ (see Figure 5).

There are a number of interesting observations to make about Figure 5. (1) It is the basis of one of the standard proofs of the Pythagorean Theorem. (2) You can *see* that s achieves its maximum when $a = b = s$; this shows immediately that, for all rectangles of a given perimeter, the square encloses the maximal area. (3) Since $s = \sqrt{ab}$, and $\frac{(a+b)}{2}$ is the radius of the circle, we have immediately that the geometric mean is always less than the arithmetic mean and is equal to it if and only if $a = b = s$.

After looking at the quadrature of some other rectilinear figures, I go on to discuss the lunes of Hippocrates **[3]** and continue with some modified versions

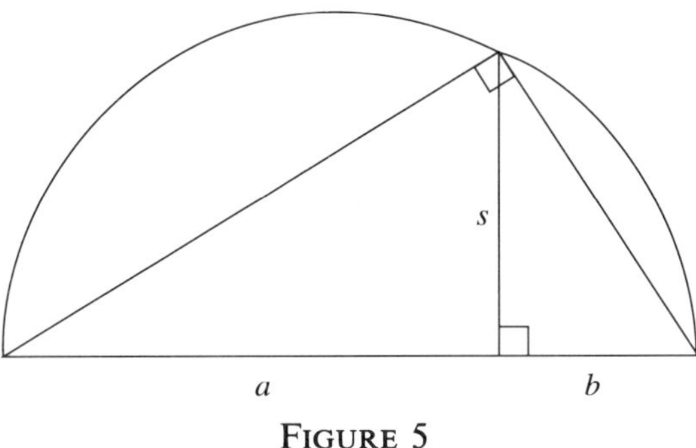

FIGURE 5

of Archimedes' quadrature of the parabolic section—including his elegant mechanical lever argument from *The Method* [4]. I continue by showing how to tackle $\int_0^a x^k\,dx$, following along the lines of Cavalieri, but using Riemann sums [2]. This method leads to the problem of finding closed formulas for the sum $1^k + 2^k + \cdots + n^k$, and so we pause to discuss these formulas. I show a variety of pictures and diagrams that give the answer for $k = 1, 2$, and 3. After that, we look at some of the recursion formulas for the higher powers of k. I then present Fermat's elegant solution [2] for evaluating $\int_0^a x^k\,dx$. Fermat completely avoids the complicated formulas above. By using an appropriate partition of $[0, a]$ into *unequal* intervals, he needs only the much easier summation formulas for the finite and infinite geometric series.

11. The Fundamental Theorem. We discuss the proof, a little history, intuitive understanding, and applications.

12. The natural logarithm and the exponential functions.

13. Solids of revolution.

14. A look back—and forward. Now, at the end of the course, is the time, it seems to me, to raise the question of why we might need more rigor in our continuing study of calculus. We look at some examples of what can go wrong when we use intuitive arguments and vague concepts like "near" or "thin" slice, etc. Here are three examples of such arguments, none completely implausible, which lead to clearly absurd answers.

(1) The step function (see Figure 6) gets "arbitrarily close" to the diagonal, but its length does not approach the length of the diagonal!

(2) There is a one-to-one correspondence between the "infinitesimally" thin slices s and s' (of equal height) in triangle ADC and triangle BDC (see Figure 7). But clearly the two triangles have different areas.

(3) The Paradox of Galileo [5]: Suppose the little circle C' is rigidly attached to the big circle C, so that it cannot slide around in any way (Figure 8). If the big circle is rolled through one revolution, it will, of course, trace out a line AB equal in length to its circumference. At the same time, the

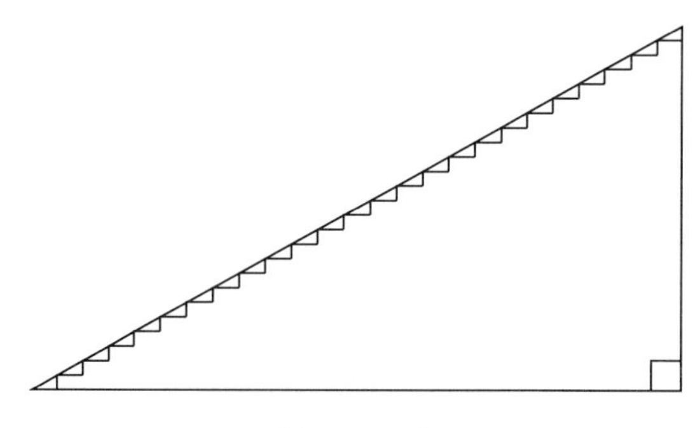

FIGURE 6

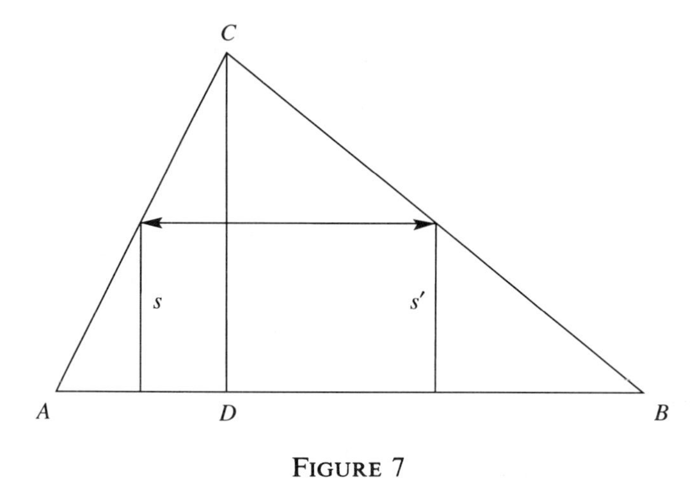

FIGURE 7

little circle will also make one revolution, but it will have traced out a length greater than its circumference! How can this be? Conversely, if the little circle is rolled through one turn it will trace out a line equal to its circumference, while the big circle traces out a line less than its circumference. How can that be?

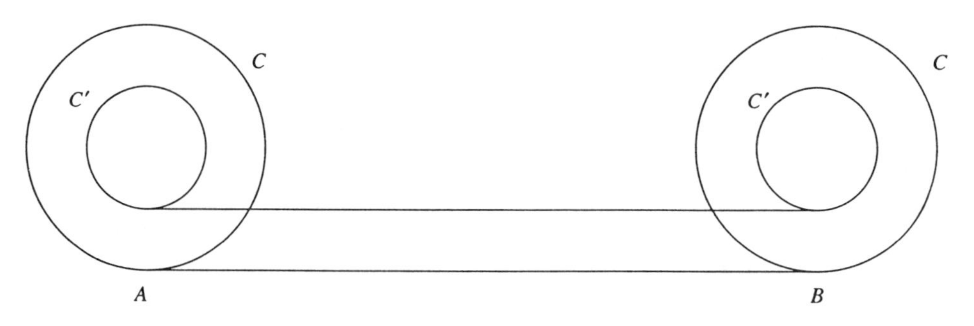

FIGURE 8

Bibliography

1. Felix Klein, *Elementary mathematics from an advanced standpoint*, Dover, New York (1960).

2. Otto Toeplitz, *The calculus; a genetic approach*, Univ. of Chicago Press, Chicago (1963).

3. Carl B. Boyer, *The history of the calculus and its conceptual development*, Dover, New York (1959).

4. Thomas Heath, *The works of Archimedes*, Dover, New York (1953).

5. Galileo Galilei, *Two new sciences*, transl. by Stillman Drake, Univ. of Wisconsin Press, Madison (1974).

National Science Foundation, Teacher Enhancement Program, 1800 G Street, N.W., Washington, D.C. 20550
E-mail address: pbraunfe@nsf.gov

CBMS Issues in Mathematics Education
Volume 3, 1993

A Program for High School Teachers on Mathematics and its Applications

DEBORAH TEPPER HAIMO

INTRODUCTION

From the time in 1983 when the National Commission on Excellence in Education issued its report *A Nation at Risk*: *The Imperative for Educational Reform*, decrying the ominous erosion of our educational system, persistent reports, particularly those generated by the National Research Council, have been keeping the issue in the forefront of public attention by providing continued substantive data that support the grimness of the situation.

We hear what many of us have long since experienced in our classrooms: our students are enrolling in our courses with ever weaker backgrounds; they are, on the whole, not interested in, nor are they studying or learning any more mathematics than they deem necessary; and they are dropping out at an alarming rate. Indeed, the statistics point to the fact that, beginning with the ninth grade and continuing through graduate school, we lose each year approximately one-half of those enrolled in all mathematics classes. At the same time, a root of the problem is highlighted by the figures that show that over half of our current secondary school mathematics teachers fail to meet even minimal professional standards for teaching the subject.

We are faced, on the one hand, with students who show no interest in a subject whose intrinsic elegance they are unprepared to recognize, and whose relevance to their everyday life they fail to comprehend. On the other, we have the problem of inadequately educated teachers who have little understanding and appreciation of mathematics as more than a discipline limited to tricky manipulations and tedious computation. It has been generally clear for some time that we must seek remedies for these very serious problems.

With teaching one of the major components of my professional life, I

This paper was a presentation made at the AMS-MAA Special Session, "Perspectives on the Roles of Research Mathematicians in Education", at the national winter meetings in San Francisco, California, on January 18, 1991.

©1993 American Mathematical Society
1047-398X/93 $1.00 + $.25 per page

could not escape the realization that my students' preparation for my courses was declining perceptibly over the years, and that the occasional high school teachers in upper-division classes were, as a rule, marginal. Beyond the effort to seek ways of addressing my own specific pressing problems, I had no thought of trying to become involved in educational issues.

Quite by chance, in the mid 1980s, I was made more immediately and directly aware of the nature and seriousness of the situation. One of my colleagues had been engaged in a preliminary development of a short summer course specifically designed to introduce high school teachers to applications in mathematics. When the project was unexpectedly funded for a period when he was otherwise committed, I was asked to substitute. The assignment was unnerving. I faced a class of teachers, all nominally satisfying a year's calculus prerequisite, but with, generally, poor understanding even of the mathematics they taught and, more surprisingly, with little facility at a computational level. It was a disheartening and revealing experience.

My initial inclination to escape from such encounters was tempered by the realization that it was essential for some of us to become involved and to take direct action to address the problems. In 1988, enlisting the cooperation of my colleague, Richard Friedlander, who has extensive expertise in teacher education, I finally decided to embark in this new direction.

An NSF Supported Program

We addressed the problem by proposing a multifaceted program aimed at enhancing the mathematical background of a select group of high school mathematics teachers and at exposing them to a wide range of applications of mathematics. The objective was to have the teachers sufficiently knowledgeable of substantive applications so that they would be better prepared to arouse their students' interest in mathematics by motivating topics through the introduction of examples that have some relevance. To this end, we felt that, above all, it was important for the teachers to have a sound understanding of mathematics and its power, so that they could explain the underlying mathematics in a meaningful way. In February 1988, we undertook the project under a National Science Foundation grant.

The program, now nearing its termination, begins each year with the selection of a group of local high school teachers. In admitting applicants to our Institute, we take into account their previous academic record. As a minimum, we require that they have at least one full year of calculus, though we have found that this may not have contributed too much to their mathematical maturity; we want them to have three or more years of experience in teaching mathematics at the high school level so that they will have sufficient confidence to experiment with the curriculum; we insist that they have the support of their school principal or some higher administrator in their school district so that their efforts are recognized and encouraged; and we ask for their commitment to participate fully in all aspects of the ten month schedule.

Over the three years, the teachers who have applied and been selected come from a variety of schools, both large and small, rural and urban, public and independent. Each successive year, our applicants include some who come from increasingly distant locations, travelling many miles to participate in our program. Their mathematical backgrounds are highly diverse, including those with recent exposure to doctoral level mathematics courses and others with limited mathematical understanding beyond the routine computational. They all share an eagerness to learn, but some have difficulty in seeing why they have to know anything that they cannot perceive as being directly applicable to their teaching. "I expected to have my mathematical knowledge increased," wrote one teacher at the program's end, concluding dolefully, "My goal has been overachieved."

The Spring Mathematics Course

To emphasize the mathematical basis of our program, our Institute begins with a challenging in-service mathematics course that meets for four half-days on successive Saturdays in the spring. It is designed to get teachers thinking mathematically about nonroutine problems; in some cases, for the first time in their educational experience. In so limited a time, we can merely introduce, at each meeting, what we feel is an interesting problem whose solution involves some important mathematical concepts that need to be stressed.

For example, we consider the question of the number of ways to build a cubic box, based on Edgar Gilbert's 1971 paper in the *Mathematics Teacher.* This problem allows us to provide for concrete experimentation; to emphasize the importance of careful, precise formulations, both of conditions assumed and results to be derived; to translate all given information into mathematical terms; to encourage conjecture; to identify and apply relevant mathematical techniques; to make some headway, but to reach a seeming impasse; and finally to explore the possibility of finding a solution by first turning to a more manageable, equivalent dual problem.

Among additional examples that we have introduced for investigation are some special magic squares leading to vector and inner product spaces, induction and Jensen's inequality, approximations and Tchebychev polynomials, and some partition identities.

To complete this component of the program, based on the examples studied in the course, we prepare some penetrating problems to be solved. We encourage the teachers to discuss them with one another freely and to help each other with difficulties. We are also available to answer questions or provide needed explanations. Solutions are to be written up individually and submitted to us at the beginning of the summer segment of the program.

In general, the teachers find this portion of the program to be very demanding and are convinced that our warning that they may find themselves "stretched" intellectually as never before is not an exaggeration. Nonetheless, most persist and work very hard, determined to extend their understanding of mathematics to a level beyond the familiar routine. Indeed, one teacher was so excited over the experience that she exclaimed enthusiastically, "Now I know why I liked mathematics particularly and why I wanted to teach that subject!"

The Summer Applications Program

On enrolling in the NSF Institute, the teachers are alerted to the fact that, following the summer program, they are to prepare a major paper on some mathematically based application. We maintain a Resource Center with a broad range of books, journals, pamphlets, and videos. Each class day, we have an extended break to allow time for serious browsing. In addition, materials in the Resource Center can be examined at any available time and can be borrowed for reasonable periods.

The formal summer schedule entails ten full-day meetings, extending over three and a half weeks. Each day's schedule centers on a visiting lecturer, invited to present an actual project that involves mathematical techniques. As a complement to each presentation, we introduce and develop the pertinent mathematics and include related applications. Among our 22 visitors, we have had a plastic surgeon who spoke on the Z- and the W-plasty; a urologist who spoke on the lithotripter; a highway patrolman who spoke on accident reconstruction; an airport engineer who spoke on the removal of asbestos from the Lambert airport ceiling; a meteorologist who spoke on weather prediction; a political scientist who spoke on the free rider problem; a food packing expert who spoke on the most economical way to pack frozen chickens; and a sociologist who spoke on designing surveys to determine a characterization of alcohol users and teenage suicides.

During the last two days of the summer course, the teachers themselves are the lecturers, each presenting an outline of a proposed applications project, with a segment at a level suitable for an actual high school class.

The completed projects are due at the end of the summer. Although deciding on a suitable topic that includes substantive mathematics beyond the routine presents some difficulties, once our teachers make a choice, they delve into it with considerable enthusiasm. On the other hand, their exposition, on the whole, leaves much to be desired. It is clear that most have never had to write a mathematics paper before, and they have a great deal of difficulty in expressing themselves clearly and in explaining mathematical ideas. Their judgement on what needs to be included and what can be assumed is rather faulty, since they often include details on trivial steps that can, and should, be done mentally, while failing to define vague terms or glossing over serious points. This is an area that needs considerably more attention.

All the projects are duplicated and distributed to the teachers so that they have some useful applications for their classes. Some of the better projects submitted to us include those on wallpaper symmetries and patterns; approximations of minimum automobile speeds by accident reconstruction; the problem of packing pizzas; the Poisson distribution as an aid to meeting federal standards of soap box contents; the Fourier series as applied to music and sound; the best use of land for planting on a family farm; and Steiner points and minimal networks.

The Fall Program

On returning to their classes in the fall, the teachers are required to send in regular reports of the specific applications they introduce into their courses each week. Once during the semester, each teacher is visited, by prearrangement, during a class session primarily devoted to an application. Opportunity is then afforded for reviewing any individual issues or resolving problems arising in the implementation of the classroom phase of the project.

During a December meeting of the group, there is a general discussion of what has or has not worked for individual teachers. Plans are also made for the culminating event of each year's Institute, the February Conference. It is an all-day affair for area-wide teachers and administrators. A featured speaker is invited to give a keynote address, and the Institute teachers give presentations of some application they want to share with their peers. There is occasion for substantial interaction among all participants. The last of these conferences will be held in February for the third and final group of teachers.

As our project comes to an end, we are scheduling a June all-day meeting with teachers from the three groups to discuss plans for their continuing role. We want them to remain in contact with one another, forming a network of peers who communicate readily and provide help and support when needed. As some have already done, we would like these teachers to hold workshops both in their schools and elsewhere to help their colleagues; we would like them to give public talks at professional meetings, both locally and nationally; and we would like them to write articles and publicize their successful experiments in teaching.

Throughout the project, we have had an outside consultant evaluate each phase of our program. The assessment is limited to its effects on the teachers, with no plans to extend the study directly to the performance of their students. The preliminary reports indicate that, thus far, the teachers who participated in our first two Institutes are still continuing to incorporate applications into their classes. They report that they rarely are asked now, "What is all this good for?" since they anticipate it as a valid question on the whole, and they seem to have greater confidence in their mathematical ability.

The program has been extremely demanding, not only on the teachers, but on the directors as well. There have been some rewarding moments, most eloquently expressed by one teacher who commented on having had to work harder than in any course before; but as the teacher noted, "The effort led to renewed faith in myself."

POSTSCRIPT—APRIL 20, 1992

The June 1991 meeting, held as planned, began with a morning session, open to all area-wide educators. They heard three invited guest speakers talk, in turn, about formation of the Missouri Coalition; available foundation, state, and federal funding for teachers; and a variety of useful mathematical applications. The speakers stayed on to participate in the afternoon session, devoted entirely to Institute teachers, some of whom presented their more successful projects on applications. There ensued a lively general discussion and a good exchange of classroom experiences with students, as well as reports of reactions of peers, and school officials. The proceedings concluded with the selection of representatives from each year's group to help maintain contact and make plans for a similar program for this June.

Over the past year, a number of teachers have taken the initiative to describe opportunities they attribute to their NSF Institute experience. Among these are several who have given presentations and held workshops for colleagues; one who decided to enroll in graduate school for further study toward an advanced degree; and another who was selected as a nominee for her state's Presidential Award for Distinguished Teaching of Mathematics. One teacher was particularly excited that a modified version of her Institute project had been published in the *Mathematics Teacher*, and this had led to her being included as a speaker at last year's national NCTM meeting and to being invited again to the next one. As one who had considered herself a good teacher but never expected to affect anyone other than the students in her assigned classes, she exemplifies the growth we aimed for in our project, and already has had substantial impact, not only in her locale, but well beyond the confines of her own classroom and region.

As recently released, Guidelines for Teacher Enhancement programs have changed substantially. The leadership program most nearly resembles our project, and it is encouraging to note that there is a requirement of a minimum of four to six weeks of instruction—a welcome improvement. Changes in pedagogical approaches are also steps in the right direction. Of at least equal importance is emphasis on sound knowledge of substantial content. Every effective means must be used to encourage teachers to extend their knowledge and understanding of mathematics far beyond the level needed for their classrooms, so that they can gain the deep appreciation of the intrinsic beauty of mathematics and of its great power for broad application, and thus be in a stronger position to arouse student interest in learning mathematics.

DEPARTMENT OF MATHEMATICS AND COMPUTER SCIENCE, UNIVERSITY OF MISSOURI-ST. LOUIS, ST. LOUIS, MISSOURI 63121
E-mail address: c1958@umslvma.bitnet

CBMS Issues in Mathematics Education
Volume 3, 1993

A Five-Year Evaluation of a Unique Graduate Program in Mathematics Education

EFFAT A. MOUSSA AND JERRY GOLDMAN

ABSTRACT. In 1984, the Department of Mathematical Sciences at DePaul University started offering a master's degree program intended for currently certified mathematics teachers interested in strengthening their command of grade 7–12 mathematics, currently or previously certified teachers in non-mathematics fields who wished to teach secondary school mathematics, or bachelor's degree holders in all fields who wished to enter teaching. The program has two unique features. The first is a content-centered curriculum tied to upper-elementary and secondary school mathematics, and the second is a two-weekend-per-month course offering structure during the academic year. The program may be completed in eighteen months. A description of the program and the results of a five-year program evaluation, with both objective and subjective components, are presented in this paper. Judged by their willingness to recommend the program to colleagues, participants were very satisfied with the program. General improvement in participants' knowledge was found in all nine areas of mathematics covered in a national test created by Educational Testing Service. Particularly striking improvement occurred in areas emphasized by the program.

1. Introduction, Program Motivation, and Needs Served

Introduction. The Master of Arts in Mathematics Education is a degree granted upon completion of a current DePaul University program, administered by the Department of Mathematical Sciences within the College of Liberal Arts and Sciences. It was conceived in 1982, first offered to students as a degree program in January 1984, then again in September 1984. It has been offered each September since then to new cohorts of students. It is an intensive 48-quarter-hour program with mainly weekend class meetings and may be completed in a year and one half. It has received generous funding from a number of outside sources, including the Amoco Foundation, Illinois Bell Telephone Company, the Illinois Board of Higher Education, and the Sherman Fairchild Foundation.

©1993 American Mathematical Society
1047-398X/93 $1.00 + $.25 per page

The evaluation of this program was considered, from the outset, as an integral part of its design. In addition to measuring whether or not the program was doing its basic job, increasing the mathematical knowledge of its participants, the evaluation was expected to provide information crucial to the evolution of the program. We think the program has done its basic job. This paper provides a portion of the basis for this belief, and we offer it as a description of one approach to others interested in developing a masters degree program to improve the competence of urban secondary school teachers.

Program motivation and needs served. An important factor in the shaping of the program was the first report [**1**, **2**] of the Commission on Precollege Education in Mathematics, Science, and Technology, a prestigious panel constituted by the National Science Board. The Commission considered the documented shortage of superior teachers to be a prime contributing cause of decreasing student participation and achievement in mathematics. Three of the specific problems for teachers which we reproduce from this report were:

1. Among certified teachers of high school mathematics and science, very few have had the formal educational preparation required to provide students with an understanding of modern technology.
2. There are few available opportunities for certified mathematics and science teachers to update or broaden their skills and backgrounds. Such training opportunities are essential due to thc rapid advances taking place in mathematics, science, and technology and the need to introduce new types of upper level courses for nonspecialists.
3. There are few in-service programs to certify teachers who are presently not qualified to teach mathematics and science.

The program we developed was intended to address the above national needs as they were reflected locally in metropolitan Chicago. In addition, we specifically hoped to address the need for students from historically underrepresented and underserved groups to have greater access to mathematics training. Thus, we wanted to make it possible, especially for participating inner-city teachers, to prepare more minority students for admission to quantitatively-oriented college programs.

The evaluation to be detailed in this paper concerns the mathematical attainments of teachers who participated in the program. The authors are currently planning a related study, which seeks to determine the ways in which program participants are applying what they have learned to their own schools.

Since the inception of the program in 1984 and through 1991, 319 students enrolled. The majority of the students were high school or junior high school teachers. A total of 224 students came from 141 different schools. The remaining 95 students were affiliated with 55 different public and private agencies or with industry. Of them, 185 students were public school teachers,

38 teachers taught in private schools, and one teacher did not identify the school type. Of all the teachers in the program, 134 were Chicago school teachers, 89 were suburban school teachers, and one teacher did not identify the school location.

2. The Program

Program description and objectives. The program is intended for currently certified mathematics teachers interested in strengthening their command of grades 7–12 mathematics, currently or previously certified teachers in non-mathematics fields who wish to change disciplines and teach secondary school mathematics, or bachelor's degree holders in all fields who wish to enter teaching. Our technical objectives reflect the needs cited in the previous section. We aim to provide the formal mathematical training necessary for understanding and teaching high school mathematics, to train teachers to become fluent in a computer language and software applicable to their classes, and to supply course work sufficient to satisfy the state of Illinois mathematics content certification requirements.

The program has two unique features: an innovative (especially in 1984) curriculum package which stresses mathematical content tied to upper-elementary and secondary school mathematics and a two-weekend-per-month course offering structure during the academic year. Whereas the program was designed to serve the mathematical needs of all Chicago area teachers, in practice we shall see it has provided an opportunity for a significant number of minority teachers in Chicago schools to earn a master's degree. It has been meeting the needs of teachers in public as well as private schools in the Chicago and suburban communities. In particular, participants in the program were queried about material covered which was most useful in their own teaching. Geometry, elementary algebra, and functions were chosen as most useful, closely followed by computer science and the history of mathematics. More specific information will be provided in §4 of this paper.

The entire degree program consists of a carefully articulated sequence of 12 courses, taken over six quarters, by a cohort of students who provide communal support to each other as they move through the curriculum together. The curriculum includes three courses in calculus, two in probability and statistics, two in history of mathematics, one in geometry, one in LOGO, one in geometry using LOGO, one in discrete structures, and one on teaching and learning secondary school mathematics. The course time offerings are given in Table 1. Typically, the time schedule for a course includes two six-hour weekend days at the beginning, two six-hour weekend days in the middle, and one six-hour weekend day at the end of a ten-week quarter. Problem sessions are held (usually before exams) in the evenings during the quarter and additional communication with instructors between class meetings takes place by mail, telephone, and personal consulting appointments. We strongly urge the students to form study groups on their own and we furnish name and address lists during the first class meetings to facilitate this. The usefulness of this peer support system is difficult to overemphasize.

TABLE 1. Course offering sequences.

	Autumn	Winter	Spring	Summer
Year I	Calculus & Analysis I	Calculus & Analysis II	Calculus & Analysis III	Teaching Secondary School Mathematics
	Geometry	History of Math. I	History of Math II	Discrete Structures
Year II	Probability & Statistics I	Probability & Statistics II		
	LOGO	Turtle Geometry		

In order to graduate, participants must have mastered basic concepts and techniques pertaining to:

1. Circular functions, analytic geometry, differential and integral calculus of one variable, certain numerical algorithms related to calculus, implementation of these algorithms using a computer language, and a software package.*
2. Euclidean and noneuclidean geometry with emphasis on Euclidean content for high school curricula.
3. The LOGO computer language and its numerical and graphical capabilities.
4. Discrete mathematical structures, modular number systems, fundamental algebraic structures, graphs, and matrices.
5. Calculus-based probability and statistics, together with an introduction to combinatorics.
6. History of elementary number theory, algebra, and geometry with particular emphasis upon studying concrete problems.
7. Methods and current research results for teaching mathematics courses.
8. Use of personal computers as class teaching tools.

Program uniqueness. We have alluded to the fact that precalculus material was regarded as the most useful by our students. We made the assumption that this content would be better learned in view of the necessity of its application to the calculus and probability sequences. It was surprising to us, in the early stages of the program, that the history sequence was a wonderful vehicle for the motivation and learning of precalculus mathematics. Naturally, we hoped this would occur while initially planning the curriculum; however, we did not foresee that a history sequence would supply our teachers with great in-class anecdotes and with many answers to the standard questions about ultimate uses of mathematics.

* This list was correct during the 1985–1989 evaluation period. Beginning in 1991, we began to stress the use of graphing calculators in calculus. In 1992, we expect to extend this to the probability and statistics sequence.

In looking for similar Chicago area programs, one finds that there are no masters degree programs in secondary teaching of mathematics offered at Northwestern, Loyola, or Roosevelt Universities. Illinois Institute of Technology has a program that exists on the books, but admission to it is suspended. The 1990/1991 University of Illinois at Chicago bulletin shows a masters program which is reminiscent of the "traditional mathematics M.S. program". The required courses, on paper, are more advanced than ours and assume a very good undergraduate preparation. In practice, this assumption is often not realistic for high school teachers. As stated, the requirements include courses in: analysis beyond advanced calculus I or II; algebra beyond linear algebra I and abstract algebra I; advanced Euclidean geometry I, II, III, and geometry beyond introduction to higher geometry; then a selection of mathematics courses from logic, finite differences, number theory, history of mathematics, topology, computer science, or probability and statistics.

Another masters program philosophy holds that education is a process which can only be mastered after the content basis is established (presumably in a demanding undergraduate program). Thus, two area schools which adhere to this approach require mostly courses in educational theory, methods, materials, and "clinical experience". For example, the University of Chicago tailors mathematics programs to individuals and does not require specific amounts of (mathematical) "content" material, but requires all of the above, and Bradley University requires at most 15 of 33 semester hours to be in "content" fields. Full time or day study is assumed in both cases.

Consequently, our curriculum, weekend structure, and completion time, taken individually or as a package, are still unique in this area.

3. Program Evaluation

Experimental method and instruments. We wanted to determine whether we were teaching our students the mathematics specified in our curriculum. Moreover, we wanted to assess the participants' degree of satisfaction with our program as measured by the extent to which they indicated their needs were met. Finally, we wanted to compile a demographic profile of the students served by the program.

Two questionnaries were designed and used for obtaining demographic, experiental, and anecdotal information from program participants. Copies of these may be found in Appendix I to this paper. The first questionnaire plus entry files were used to gather demographic data, work experience, reasons for enrollment in the program, and other background data from entering cohorts. The second questionnaire was completed by participants upon completion of the program. This final questionnaire requested information about reactions to the program in terms of knowledge gained, satisfaction of needs, as well as suggestions for improvement.

A review of the test and measurement literature indicated that no single instrument existed which would precisely measure the mathematical knowledge impact of our particular curriculum upon participants. Beginning in June 1985, we began a formal objective evaluation scheme for our program with the cooperation of Educational Testing Service (ETS) of Princeton, New Jersey. A form of the Mathematics Specialty Area test of the National Teacher Exam (NTE) had a reasonable degree of concordance with our program and was as close to a single instrument as we were likely to come. There was no possibility of a large percentage of our students agreeing to sit through a battery of tests.

We decided to use this NTE test as a portion of our evaluation because it was a reasonable existing objective measure, because it allowed us to compare our students with national norms, because of the expertise of ETS in reliable test construction, and finally, because the act of sitting for it provided good experience for our participants. Educational Testing Service, through its NTE field representative in Evanston, Illinois, very kindly agreed to cooperate with us by furnishing, administering, and scoring the exams. No record of participants' names or scores was kept by ETS, and only aggregated data will be released by DePaul.

The questions on the test cover two major areas:

1. Mathematical content ranging from arithmetic and algebra to AP calculus, abstract and linear algebra, finite mathematics, number theory, and probability and statistics;

2. Mathematics education, emphasizing pedagogy but including knowledge of professional organizations and journals, current reports, trends and curricular development, and the history of mathematics.

In addition, as ETS points out, the examination attempts to measure the ability to "interpret symbols and terms, demonstrate skills, apply concepts, extend concepts to unfamiliar situations, justify statements and construct proofs, relate essential knowledge and concepts to a learning situation, select appropriate teaching methods and principles, and identify and interpret trends and curriculum development in the light of the history of mathematics education" [**3**].

The National Teacher Examination Specialty Area Test in Mathematics is not often used for program evaluation. The exam is designed to measure the mathematical knowledge and abilities of examinees who plan to teach the subject at the secondary level and it might typically be used as part of a certification process. Its validity is based ultimately upon its content specifications, which have been reviewed by various experts (public school teachers and administrators, teacher educators, and subject matter specialists) across the country [**3**]. The test has 120 questions covering nine content categories, which have a large intersection with the DePaul program curriculum content. However, the fit is not perfect. Table 2 gives the nine areas on the test and the percentages of the test items. Test items which are not formally covered in our program are those content items in transformation geometry and in linear algebra in categories III and VI. On the other hand, certain items within the curriculum are not covered on the examination. Examples of these are univariate and jointly distributed random variables, theory of estimation and hypothesis testing in the area of probability and statistics, and the use of computer software for simulation and data analysis as well as graphing packages and calculators in calculus. There are also some differences in emphasis. For example, probability and statistics (category V) accounts for 6% of the NTE Mathematics exam, whereas these content items are covered in two courses out of 12, or 17% of our program.

The ETS/NTE test was administered to program entrants in 1985 as a pre-test; then it was administered again, one and one half years later, as a post-test to those members of the entering cohort who had completed the program. This process was repeated for a total of five cohorts, 1985–1989. The same test was given, no tests were retained by takers, and ETS handled the security.

Scores on the number of correctly answered questions in each of the nine areas, the total score on the test, and a scaled score * obtained by ETS from the raw score of each student are used in the analysis.

4. Results of the Evaluation

Demographic and self-assessment data. The five-year evaluation study (1985–1989) included 128 students. Of them, 95 (74.2%) had graduated by the year 1991, nine (7%) are still actively enrolled in the program, and 24 (18.8%) have dropped out of the program. There were 90 (70.3%) females and 38 (29.7%) males in the study. Eighty-one students (63.5%) were under the age of 40, and 7 students (5.5%) were 50 years of age or older. Recall that other data regarding the teaching level of our student population was given at the end of Section 1.

Eighty students (62.5%) were white, 37 (28.9%) were African-Americans, three (2.3%) were Hispanic Americans, and eight (6.3%) were from other ethnic groups. Of those students who responded, about 60% had mathematics degrees, and 40% had nonmathematics degrees. There were 107 students (87%) with bachelor's degrees, and 16 (13%) with master's degrees. Ninety students responded to the question in the first questionnaire of Appendix I about their number of publications. Of those who responded, 83 students (92.2%) had no publications, six students (6.7%) had one, and one student had three or more publications.

Of all the students who responded to the question in the first questionnaire regarding their teaching assignments, seven students (5.7%) were not teaching, 107 students (87%) were teaching mathematics, and nine students (7.3%) were teaching other subjects. When asked about the number of years

* See Appendix II for the definition of "scaled score" and for Table 3, which provides national scaled score statistics.

TABLE 2. Distribution of ETS test questions by category.

Category	Category Content	# of Items	Percent
I	Number Concepts, Elementary Number Theory	8	6.7
II	Elementary and Intermediate Algebra	14	11.6
III	Geometry—including Euclidean Solid, Coordinate, and Transformational	12	10.0
IV	Functions and their graphs—including Algebraic, Trigonometry, Logarithmic, and Exponential	20	16.7
V	Counting Problems, Probability, and Statistics	7	5.8
VI	Algebraic topics—Structural Systems, Abstract and Linear Algebra	14	11.7
VII	Limits, Calculus, and Properties of Real Numbers	18	15.0
VIII	Other topics: Measurement, Computer Science, Finite Math, and History of Mathematics	11	9.2
IX	Professional Knowledge	16	13.3
Total		120	100

of teaching, the responses ranged from zero for those who did not teach to 30 years for those who did. Forty-eight of the respondents (39.3%) had under five years of teaching experience, and 24 students (19.7%) had at least sixteen years of experience. The mean number of years of teaching was 8.975 years, the standard deviation was 7.096 years, the median was seven years, and the mode was two years.

In summary, the majority of the students in this study were white females, under age 40, with bachelor's degrees in mathematics, teaching mathematics in high or middle schools, with under ten years of experience on their jobs, and with no publications.

Students were asked to select one or more reasons for enrolling in the program. 'Increasing mathematical knowledge' was the reason chosen with the highest frequency, selected by 107 (87%) of the respondents, followed by 'improving teaching abilities in mathematics', selected by 104 students (84.6%), and 'increasing salary level', which was selected by 72 students (59%). Other reasons chosen were 'retraining in mathematics', selected by 44 students (35.8%), and 'possible change in position', selected by 41 students (31.5%).

When asked to rank their job satisfaction level on a scale of one (low) to five (very high), 106 students (82.8%) indicated a ranking of 3 or more. The mean rank was 3.697, with the median and mode equal to four.

The following question was posed: "If you are teaching mathematics, do you believe that enrolling in the present program will improve your teaching skills in ... ". A list of areas of expected improvement was then given. Seventy-eight of the respondents (64.5%) selected the area of geometry alone, while 15 students (12.4%) chose it with other areas. Algebra was selected alone by 80 students (66.1%), while 15 students (12.4%) chose algebra with other areas. Forty-two students selected computer science alone (34.7%), and 15 students (12.4%) chose computer science with other areas. Other areas specified by the respondents were calculus, precalculus and analytic geometry, trigonometry, and preparing junior high school students for high school education.

As previously indicated, a second questionnaire was distributed toward the end of the program, before graduation. This final questionnaire was distributed to help in evaluating participants' experiences relating to the program. Of those who responded to the question regarding their teaching assignment (cf. Appendix I), 67 students (91.8%) indicated that they were teaching mathematics, while six students (8.2%) were science teachers. Fifty-five students did not respond to this question.

Students were asked to list the areas in which they felt they lacked knowledge at the time they took the ETS pre-test. The nine areas shown in Table 2 were listed and students could choose one or more of them. Table 4 gives the percentages of students who chose each of the nine areas. Probability and statistics was the most frequent response chosen by 44 (69.8%) of the students who answered this question. Over 55% of the responding students indicated lack of knowledge in each of calculus, computer science, and history of mathematics.

When asked the question, "After taking courses at DePaul, in which of the following areas do you feel your knowledge has increased?", again, the area of probability and statistics was predominantly selected, followed by the areas of calculus, computer science, and history of mathematics. The respective percentages of responding students were 86.8%, 80.3%, and 78.9% in these areas. Table 5 gives the percentages for all nine areas.

Asked about the usefulness of the program to their teaching assignment, 66 students (88% of those who responded to the question) felt it was very useful, and 53 did not answer the question. Furthermore, students were asked to rank all areas from most useful (1) to least useful (9). The mean ranks given to each area were computed and are given in Table 6. Geometry, elementary algebra, and functions were ranked as most useful, having mean ranks of 3.0 or less. Probability and statistics had a mean rank of 3.8, while abstract and linear algebra had the highest mean rank of 4.3, which is still of moderate usefulness in their teaching.

When asked to rate the program of study of DePaul on a scale of one (poor) to five (excellent), 81.3% of those who responded to the question gave it the rating of very good to excellent. Only two (1.3%) gave it a rating of 2 (average), and none gave it a poor rating. In addition, in response to the question, "Would you recommend this program to other colleagues?", *all* the students answered in the affirmative.

In summary, the students entering the program perceived themselves as lacking mathematical knowledge in the areas identified by the program and were seeking to increase their knowledge in these areas. According to the students' self-evaluations, students increased their knowledge in most of the areas covered in the program and found the program useful in their classroom teaching. Students' indication that they would recommend the program to their colleagues is interpreted as reflecting their satisfaction with the program.

TABLE 4. Distribution of respondents by areas of self-indicated lack of knowledge at the time of the ETS pre-test.

Area	Frequency	Percentage	Valid Percentage*
Probability and Statistics	44	34.4	69.8
Calculus	35	27.3	55.6
CSC and History of Math	35	27.3	55.6
Abstract & Linear Algebra	30	23.4	47.6
Number Theory	25	19.5	39.7
Professional Knowledge	21	16.4	33.3
Functions	17	13.3	27.0
Geometry	14	10.9	22.2
Elementary Algebra	1	0.8	1.6

* 63 valid responses (among the 128 in the study).

Results of the national teacher examination in mathematics. Not all the students who took the ETS/NTE pre-test took the post-test. Of the 128 students included in our five-year study, 127 (99.2%) took the pre-test, and 86 (67.2%) took the post-test. There are many reasons for such a large difference. Conflicts in scheduling the ETS exams with the teachers' schedules, dropout cases (24 cases) and delayed graduation of students who take fewer than two courses per quarter (9 cases) are some of these.

Of the 127 students who took the pre-test, 63 (49.6%) had a total score in

TABLE 5. Distribution of respondents by areas of self-indicated increased knowledge after the program.

Area	Frequency	Percent	Valid %*
Probability and Statistics	66	51.6	86.8
Calculus	61	47.7	80.3
CSC and History of Math	60	46.9	78.9
Number Theory	43	33.6	56.6
Professional Knowledge	36	28.1	47.4
Geometry	35	27.3	46.1
Functions	35	27.3	46.1
Elementary Algebra	20	15.6	26.3
Abstract and Linear Algebra	16	12.5	21.1

* 76 valid responses (among the 128 in the study).

TABLE 6. Mean rankings of areas of usefulness in teaching assignments.

Area	Mean Ranking*
Geometry	2.9
Functions	3.0
Elementary Algebra	3.0
CSC & History of Math	3.1
Calculus	3.4
Number Theory	3.6
Professional Knowledge	3.6
Probability and Statistics	3.8
Abstract and Linear Algebra	4.3

* Areas ranked from most useful (1) to least useful (9). Valid cases 74.

all nine areas of less than 60 out of 120. The corresponding figure for those who took the post-test is 19 (22%). At the other end, 17 students (13.4%) correctly answered 86 questions or more on the pre-test, as compared to 32 (37.2%) on the post-test. See Table 7 for more details. The mean number of correct questions answered was 60.362 on the pre-test, and 75.488 on the post-test. The medians were 60 and 77.5, the modal values were 76 and 89, and the standard deviations were 21.37 and 19.63, respectively. The observed difference was statistically significant ($P < 0.001$). See Tables 7–10. Although the improvements in mean scores from the pre-test to the post-test were statistically significant in all the areas (see Tables 7–10), the improvements in geometry, probability and statistics, calculus, and computer science were particularly outstanding. This may be partially explained by the fact that our program consists of three courses in calculus, two courses in geometry, two in probability and statistics, two in history of mathematics, two in discrete structures and LOGO, and one in methods of teaching secondary school mathematics.

In addition to the analysis of the number of correctly answered questions on the pre- and post-tests, the scaled scores, computed by ETS as described earlier, were compared with the national percentiles based upon the scores of 4751 mathematics examinees during the period 1982–1985. Table 11 gives the percentage distributions of the scaled scores of the participants in the five-year evaluation study, and their approximate national percentile ranks. As can be seen, 42% of those who took the pre-test had a percentile rank of 51 or lower. The corresponding percentage was 22.5% for the post-test. On the other hand, 8% of the scaled scores on the pre-test had a percentile rank of 98 or higher, compared with 20.3% on the post-test. Overall, 32% of those who took the pre-test scored at or above the national 88th percentile, while the corresponding figure was 52.5% for the post-test. The mean scaled score was 618 (which is ranked as the national 66th percentile) on the pre-test and 672 (which is ranked as the national 86th percentile) on the post-test. These may be compared with the national mean of 582 (which is ranked less than the national 51st percentile, which was 590). The standard deviations were 95, 91, and 82, respectively. More interesting are the median scaled scores, which were 620 on the pre-test (ranking as the national 66th percentile) and 685 on the post-test (ranking as the national 89th percentile). The modal values were 690 on the pre-test (ranking as the national 90th percentile) and 740 on the post-test (ranking as the national 96th percentile). See Tables 11 and 12.

5. Summary and Conclusions

Since the program's inception in 1984, the Master's Degree program in Mathematics Education has reached over 300 teachers and nonteachers in Chicago and Suburban communities. A significant number of these were secondary school teachers in public and private schools. Over 37% of the students were from historically underrepresented and underserved racial or ethnic groups, and about 70% of the students were female.

Two types of instruments were used to evaluate the program's impact on the students: self-reporting questionnaires and the National Teacher Examination Specialty Area Test in Mathematics. The self-reporting questionnaires were designed to assess: (1) students' perceived mathematical knowledge when they entered the program; (2) students' self-evaluations of what they learned in the program; (3) the program's relevance to actual teaching assignments; and (4) students' overall satisfaction with the program as judged by their willingness to recommend the program to their colleagues. The National Teacher Examination Specialty Area Test in Mathematics was used as a pre-test and post-test to evaluate the effect of the program upon their mathematical knowledge.

Significant improvement in knowledge was measured in all nine areas of mathematics covered in the ETS test. More striking improvement was found in the areas emphasized in the program, namely calculus, probability and statistics, geometry, computer science, and history of mathematics. Percentile ranks of the scaled scores on the tests were much higher than the national average, with significantly more cases in the post-test scoring over the 97th percentile than in the pre-test.

In the student self-reports, students generally indicated that at the time they entered the program, they were lacking mathematical knowledge in the areas identified by the program. In evaluating the impact of the program, the students reported that the program was useful in their classroom teaching and expressed willingness to recommend the program to their colleagues.

We think we have developed a program package that performs its basic job of increasing the mathematical competence of our urban high school teachers. In a followup study currently being planned, we expect to investigate the uses to which our participants put their training within their own schools and the impact upon their high school students. We hope this evaluation is of use to other mathematicians interested in program development.

6. Acknowledgments

The authors wish to express their appreciation to the following people for their assistance with the project, in varying degrees, at its different stages.

Special thanks go to Professor Richard J. Meister, Dean of the College of Liberal Arts and Science, DePaul University, for his continued support for the program since its inception. Moreover, we particularly wish to thank Ms. Ines Bosworth of the Educational Testing Service, Evanston, Illinois, for her valuable cooperation with the furnishing and administration of the ETS/NTE examinations, and to Mr. Phil Harvey of ETS for scoring them.

We are indebted to Mr. Joe Baker, of the Assessment Center at DePaul University, for his valuable assistance with the computer and programming aspects of the data handling. Special thanks are due the following students: Tim Dorsey, Debbie Goldman, Mark Jackowiak, and Don Opitz for their contributions in the data entry and handling stages. Last, but not least, we wish to thank the many colleagues whose dedication to teaching in, and administration of, the program made it the success that it is.

TABLE 7. Distribution of the total correctly answered questions on the ETS pre- and post-tests (not-paired).

Score	Pretest Frequency	Pretest Percent	Pretest Valid %	Post Frequency	Post Percent	Post Valid %
20–36	22	17.2	17.3	2	1.6	2.3
37–49	22	17.2	17.3	7	5.5	8.1
50–59	19	14.8	15.0	10	7.8	11.6
60–74	25	19.5	19.7	19	14.8	22.1
75–85	22	17.2	17.3	16	12.5	18.6
86–100	13	10.2	10.2	25	19.5	29.1
101–120	4	3.1	3.2	7	5.5	8.1
No Resp.	1	0.8		42	32.8	

TABLE 8. Summary statistics of the ETS pre- and post-tests.

Area	Test*	# of Ques	Mean	Med.	Mode	S	S(X)	Range
Number Theory	Pre	8	6.079	6.0	6.0	1.56	0.138	8.0
	Post	8	6.907	7.0	8.0	1.16	0.126	6.0
Elementary Algebra	Pre	14	10.669	11.0	11.0	2.57	0.228	10.0
	Post	14	11.663	12.0	13.0	1.84	0.199	9.0
Geometry	Pre	12	6.393	7.0	5.0	2.84	0.25	11.0
	Pos	12	7.674	8.0	7.0	2.55	0.275	11.0
Functions	Pre	20	9.433	9.0	7.0	5.24	0.465	19.0
	Post	20	12.012	13.0	13.0	4.72	0.509	19.0
Probability and Statistics	Pre	7	2.976	3.0	1.0	1.88	0.167	7.0
	Post	7	4.395	5.0	5.0	1.80	0.194	7.0
Abstract and Linear Algebra	Pre	14	4.598	4.0	3.0	2.41	0.214	11.0
	Post	14	6.291	6.0	6.0	2.66	0.287	12.0
Calculus	Pre	18	7.449	7.0	4.0	4.07	0.361	16.0
	Post	18	11.360	12.0	11.0	3.67	0.396	14.0
CSC and History of Math	Pre	11	4.913	5.0	5.0	2.19	0.194	10.0
	Post	11	6.069	6.0	6.0	2.23	0.232	10.0
Professional Knowledge	Pre	16	7.764	8.0	8.0	3.24	0.297	14.0
	Post	16	9.105	9.0	8.0	3.05	0.329	13.0
Total	Pre	120	60.362	60.5	76.0	21.37	1.90	86.0
	Post	120	75.488	77.5	89.0	19.63	2.12	88.0

* Figures in this table are based on 127 cases for the pre-test and 86 cases for the post-test.

TABLE 9. Paired T-tests and other statistics.

Area	Test	# Cases	Mean	S	$S(X)$	Difference	T	P *	r	P **
I	Pre	86	6.37	1.39	0.15					
						0.5349	3.67	< 0.0010	0.451	< 0.001
	Post	86	6.91	1.16	0.13					
II	Pre	86	11.29	2.32	0.25					
						0.3721	0.13	< 0.018	0.719	< 0.001
	Post	86	11.63	1.84	0.20					
III	Pre	86	7.06	2.64	0.28					
						0.6163	2.90	< 0.002	0.712	< 0.001
	Post	86	7.67	2.55	0.27					
IV	Pre	86	10.83	4.90	0.53					
						1.174	3.91	< 0.001	0.833	< 0.001
	Post	86	12.01	4.72	0.51					
V	Pre	86	3.32	1.74	0.19					
						1.070	7.38	< 0.001	0.711	< 0.001
	Post	86	4.40	1.80	0.19					
VI	Pre	86	5.02	2.49	0.27					
						1.267	6.13	< 0.001	0.725	< 0.001
	Post	86	6.29	2.66	0.28					
VII	Pre	86	8.52	3.83	0.41					
						2.837	10.3	< 1.001	0.768	< 0.001
	Post	86	11.36	3.67	0.40					
VIII	Pre	86	5.33	2.13	0.23					
						0.744	3.39	0.0005	0.565	< 0.001
	Post	86	6.07	2.23	0.24					
IX	Pre	86	8.29	3.18	0.34					
						0.814	2.77	0.003	0.619	< 0.001
	Post	86	9.10	3.05	0.33					
Tot.	Pre	86	66.16	19.80	2.14					
						9.33	9.50	< 0.001	0.893	< 0.001
	Post	86	75.49	19.63	2.12					

* P-value is one-tail probability.

** P-value is two-tail probability. Areas consist of the categories in Table 2.

TABLE 10. Comparison of the percentage* performance on the pre- and post-tests by area.

	At most half the questions (A)			At least k_i questions (B)		
Area	# of Questions n_i	Pre %	Post %	# Correct k_i	Pre %	Post %
I	8	16.4	0.8	7	44.1	67.4
II	14	11.0	2.3	12	41.7	61.6
III	12	48.8	26.7	10	17.3	26.7
IV	20	58.3	32.6	17	9.4	18.6
V	7	62.2	30.2	6	11.0	31.4
VI	14	86.6	68.6	10	4.7	12.8
VII	18	69.3	32.6	15	6.3	23.3
VIII	11	65.4	39.5	9	6.3	12.8
IX	16	61.4	41.9	13	6.3	14.0
Total	120	49.6	22.0	86	13.4	37.2

* The given percentages are the percentages of respondents who correctly answered:

(A) at most half of the questions in an area; or

(B) at least k_i out of n_i questions in the ith area, $i = 1, 2, \dots, 9$.

Areas consist of the categories of Table 2.

TABLE 11. Frequency distribution of the scaled scores on the ETS pre- and post-test by National Percentile Ranks.

		Pre-Test			Post-Test		
Scaled Scores	National Perc.* Ranks	Frequency	%	Valid %	Frequency	%	Valid %
400-440	1–3	4	3.1	4	0	0	0
450–490	4–13	7	5.5	7	2	1.6	2.4
500–540	15–29	13	10.1	13	3	2.3	3.5
550–570	34–41	13	10.2	13	11	8.6	13.1
580–590	46–51	5	3.9	5	3	2.3	3.5
600–620	57–66	11	8.6	11	11	8.6	13.1
630–640	70–75	6	4.7	6	3	2.3	3.6
650–670	78–85	9	7.0	9	7	5.5	8.3
680–690	88–90	10	7.8	10	5	3.9	6.0
700–730	91–95	10	7.8	10	14	10.9	16.7
740–750	96–97	4	3.1	4	8	6.3	9.5
760–780	98	3	2.3	3	9	7.0	10.7
790–810	99	5	3.9	5	4	3.1	4.8
820–860	99	0	0	0	4	3.1	4.8
no response		28	21.9		44	34.8	
Total		128	100	100	128	100	100

* Percentile Ranks from Table 1, "Specialty Area Test Interpretive Data-Percentile Ranks used on Score Reports", Publication H2, Educational Testing Service, 1985.

TABLE 12. Summary statistics of the percentile ranks of the scaled scores.

Statistics	National*	Pre-Test	Nat. Perc. Rank	Post-Test	Nat. Perc. Rank
Mean	582	618	66	672	86
Median		620	66	685	89
Mode		690	90	740	96
Standard deviation	82	95		91	
Smallest		400	1	450	4
Largest		810	99	860	99

* Based on a sample of 4751 cases.

APPENDIX I

Questionnaire cover sheet. The purpose of this questionnaire is to provide us with certain background information on participants in the M. A. degree program in Mathematics Education.

The information will be held in utmost confidentiality and will only be used collectively and in aggregate form, for the purpose of evaluating the program now and in future years.

DePaulUniversity

Department of the Mathematical Sciences
Master of Arts in Mathematics Education Program

2323 North Seminary Avenue
Chicago, Illinois 60614-3298

312 341-8250

Background Questionnaire

Math-Education Program

1. Name: ______________________________

2. Date of enrollment in program: Quarter ____________ Year _______

3. Sex: Male _____ Female _____

4. Age:
   Under 30 ______
   30 - 40 ______
   40 - 50 ______
   50 and over _______

5. Education:
   Bachelor — Major: Math _______; Other, specify _______
   Higher degrees? — Math. _______; Other, specify _______
   Non-degree Graduate Courses — Math. _______; Other, specify _______

6. Work Experience:

   A) Teaching: Math ____ No. of years ____ Grades taught _____
   Other, specify ____ No. of years ____ Grades taught _____

   B) Other experience: Specify ________________ No. of years ________

7. Reasons for enrolling in the Program - (You may choose more than one)

   ______Retraining in Math.
   ______Increase in salary level
   ______Increase Math. knowledge
   ______Possible change in position
   ______Improve teaching abilities in Math.
   ______Other, specify

8. On a scale of 1 to 5, please rank your job satisfaction level

Very high	High	Average	Below average	Low
5	4	3	2	1

9. If you are presently teaching Math., do you believe that enrolling in the present program will improve your teaching skills in (check one or more):

   ______ Geometry
   ______ Algebra
   ______ Computer Science
   ______ Other, specify ________________

Part I - Questionnaire

To Be Extracted from Records

1. Name ______________________________

2. Education

   Majors

   Degrees earned

3. Ethnic Background ______________________

4. Positions in Professional Organizations:

   ____

   ____

   ____

5. Research or Publications: Yes ____ No ____ If so, the # ___

6. Educational Objectives: ______________________

____

____

____

DEPAUL UNIVERSITY

MASTER OF ARTS IN MATHEMATICS EDUCATION PROGRAM

QUESTIONNAIRE FOR JUNE GRADUATES

This is your last progress follow-up questionnaire. It is part of a long-term effort to evaluate the M.A. in Mathematics Education Program. You have been enrolled in the program for about eighteen months. Your candid opinion regarding your experience is greatly appreciated.

1. Name: ____

2. In what area is your present teaching assignment?

   Math______ Science______ Other______

3. Did you take the "ETS" Mathematics Exam before enrolling in this program? Yes______ No______

4. If your answer to question #3 is <u>No</u>, please go to question #6. If your answer to question #3 is Yes, how do you rate your performance on the "ETS" exam?

   Poor______ Good______ Excellent______

5. At the time of taking the "ETS" exam, in which of the following areas did you feel you lacked knowledge?

   ____a) Number Theory
   ____b) Elementary Algebra
   ____c) Functions
   ____d) Geometry
   ____e) Probability & Statistics
   ____f) Abstract & Linear Algebra
   ____g) Calculus
   ____h) Computer Science, History of Mathematics
   ____i) Professional Knowledge

6. After taking courses at DePaul, in which of the following areas do you feel your knowledge has increased?

____a) Number Theory
____b) Elementary Algebra
____c) Functions
____d) Geometry
____e) Probability & Statistics
____f) Abstract & Linear Algebra
____g) Calculus
____h) Computer Science, History of Mathematics
____i) Professional Knowledge

7. During your time at DePaul, did you find the classes you took to be useful in your teaching assignments?

Yes______ No________

8. If your answer to question #7 is Yes, which of the following areas did you most use in your teaching? [Rank all areas from most useful (1) to the least useful (9).]

____a) Discrete Structures; Number Theory
____b) Elementary Algebra
____c) Functions
____d) Geometry
____e) Probability and Statistics
____f) Abstract & Linear Algebra
____g) Calculus
____h) Computer Science; History of Mathematics
____i) Professional Knowledge

9. If your answer to question #7 is No, please indicate the area or areas that you would have needed to use most at work, and weren't offered during your time here.

____a) Discrete Structures; Number Theory
____b) Elementary Algebra
____c) Functions
____d) Geometry
____e) Probability and Statistics
____f) Abstract & Linear Algebra
____g) Calculus
____h) Computer Science; History of Mathematics
____i) Professional Knowledge

10. How do you rate the program of study at DePaul?

Poor_____ Average_____ Good_____ Very Good____ Excellent____

11. Would you recommend this program to other colleagues?

Yes______ No______

Reasons:____

12. Do you have any suggestions or comments regarding the program? (Feel free to use the reverse side.)

GOLDMAN2:question.nai

Appendix II

The scaled scores of the ETS exams are derived from the following formula:

$S =$ a preliminary or raw score;

$C =$ the number of correctly answered questions on the test; and

$W =$ the number of wrong answers on the test.

Then, $S = C - 0.25\, W$.

S is then converted to a "scaled score", Y, in the interval [250, 990], by applying a linear map given by

$$Y = 387.7896 + 4.1284\, S$$

National interpretive data, based upon scores of 4751 mathematics examinees during a 1982–1985 time period have produced the percentile ranks given in Table 3 of "scaled scores" for examinees who were seniors in college or had a bachelor's degree.

Table 3. National statistics on the ETS/NTE scaled scores.

Scaled Score	Percentile Rank
790–990	99
780	98
640	75
590	51
530	25
410	1

The mean scaled score was 582, and the standard deviation was 82.

References

1. *Education commission to explore improvement in science, mathematics achievement*, National Science Foundation News, NSF PR82-31, 1982.

2. *Today's problems, tomorrow's crises*, A report of the National Science Board Commission on Precollege Education in Mathematics, Science and Technology, National Science Board, National Science Foundation, 1983.

3. Interpreting the NTE Speciality Area Test Scores, 1985–86, publication H2, Educational Testing Service, 1985.

Department of Mathematical Sciences, DePaul University, Chicago, Illinois 60614

E-mail address: matem@depaul.bitnet
matem@orion.depaul.edu

E-mail address: matjig@depaul.bitnet
matjig@orion.depaul.edu

CBMS Issues in Mathematics Education
Volume 3, 1993

Creating a New College Preparatory Math Course: An Overview

ELAINE KASIMATIS AND TOM SALLEE

INTRODUCTION

The College Preparatory Mathematics: Change from Within (CPM) Project of the University of California at Davis is designed to replace the traditional three-year college preparatory mathematics program of Algebra I, Geometry, and Algebra II with one that better meets the needs of students, particularly those students from groups who are historically underrepresented in mathematics, in an increasingly technological society. Funded for five years with Eisenhower monies and administered by the California Postsecondary Education Commission, the project was originally conceived as a way of teaching concepts and developing skills in a somewhat different order than is done traditionally, eliminating some material that has been outdated by modern technology and replacing it with more useful material. As implemented, however, the project has evolved so that the most important difference between our program and the traditional one is *how* the material is presented rather than *what* is presented. This article sketches how the program was conceived and why it developed along the lines that it did.

The group developing the CPM materials numbers about forty. Three directors, Elaine Kasimatis, Judy Kysh, and Tom Sallee, have overall charge of the implementation of the project, while the remainder are secondary teachers. During the spring of 1989, three meetings were held with a group of thirty teachers to discuss which topics and skills should be included in a three-year college preparatory math sequence. The ultimate decisions were heavily influenced by the *Standards* [**1**] publication of the NCTM, the *Mathematics Framework* [**3**] of the state of California, and the report *Everybody Counts* [**2**].

An early agreement was that, subject to constraints to be discussed later, the three courses should be considered as parts of a single three-year course. Further discussions produced an agreement on the goals desired at the end

©1993 American Mathematical Society
1047-398X/93 $1.00 + $.25 per page

of the three years. After coming to a general consensus on what topics and skills should be included, the group agreed which of the topics should be presented in the first year of the program. The directors then organized the sequence of units for Math I, our replacement Algebra I course, and summarized the topics to be included in each unit. During the month of July, the thirty teachers and directors wrote the units for Math 1. A small group of teachers spent another week writing preliminary teacher notes. The course was piloted by the developers during the 1989–90 academic year, then extensively rewritten during the summer of 1990. The revised materials were then used and critiqued by some of the original contributors, as well as some 30 additional teachers who were not involved in their development. The materials underwent another revision, including a polishing of the teacher notes, in summer 1991.

Our replacement geometry course, Math 2, was developed in a similar manner during the academic year 1989–90 and written during the following summer. The materials were piloted by the teacher-writers in about 40 classes during the 1990–91 academic year. Extensive revisions, which include enhanced teacher notes, were completed by a small group of developer-piloters during July and part of August, 1991.

We initially intended to develop Math 3, our replacement Algebra 2 course, during the 1990–91 academic year and write it during summer 1991. However, after experiencing the time and energy demands imposed during summer 1990 by the revision of Math 1 and the writing of Math 2, we decided it would be wise to postpone the development of Math 3 for one year and focus on the implementation and evaluation of Math 1 and 2 in the 1990–91 cycle.

Of fundamental importance to the project was the directors' strong sense of the skills and knowledge students should have when they arrive at college. On the other hand, we directors have little current teaching experience in secondary schools. Thus we needed experts who know intimately what our target students are like today and how best to reach them. These experts are the teachers who have been involved in the development and writing of the materials.

During 1990–91, 71 teachers and about 4000 students participated in the project. Both of these numbers were larger than we wanted, certainly more than we anticipated, at this stage. We are funded to work with three districts in the greater Sacramento area: Sacramento City, Elk Grove, and San Juan. In addition to the obvious goal of developing the new course, a secondary but still major goal of the project is to see if serious change is possible on a scale of thousands of students given adequate resources. We originally proposed to replace 85% of the existing mathematics courses in these districts with our program. Two years into the project, it appears that a more realistic scenario is that most of the schools in the target districts will convert completely to the CPM program while some schools will not be involved at all.

Assumptions Guiding Our Course Development

As mathematicians, we like to be as clear as possible about our assumptions. Ultimately, we made two kinds of assumptions. The first kind are the constraints we see imposed by the reality of the world in which we expect this curriculum to be implemented. Most of these assumptions grew out of conversations with teachers at an early stage of the project. Assuming that our teachers are correct, this means that these first constraints are universal—that is, they must be dealt with by any group attempting to implement a new curriculum. The second set of assumptions represents our beliefs about how best to bring about true understanding of important mathematics with students.

Assumptions about the world. Curriculum change is not only an educational process, but a political one as well. And, like politics, curriculum reform is an art of the possible. Because we see our course as a transition course to the next generation of secondary mathematics courses, and because we want it to be possible for our materials to be adopted widely, we realized that we could not immediately implement all of our ideas for reform in secondary mathematics. Rather, we needed some sense of which of our innovations would be acceptable for a significant fraction of secondary schools. To gain this sense, we listened carefully to what our teacher experts told us about the constraints on change.

The primary constraint is that *the new courses must coexist with existing courses.* No large school district can be expected to replace all of its old courses with new untried materials. Within each school involved in our program, usually two or three teachers use our materials while the remainder of the math department continues to teach the school's standard courses. This is done with the understanding that students might go into a traditional geometry course following CPM Math 1. Alternatively, students who pass a traditional Algebra I course may enter our Math 2 course the following year. As a consequence, if students are not to be disadvantaged, there has to be substantial overlap in the subject matter of a CPM course and the traditional course it replaces. We would prefer to design the CPM program so that the subject matter is more thoroughly integrated throughout the three-year sequence. However, given this constraint, we have probably done as much integration as possible.

A second major constraint is that *a new course cannot require extra work from the teachers in the long run.* In the short run, most teachers are willing to commit extra time and effort to a course for one year or possibly two. However, they are not willing to adopt a course that will require extra work every year. We have been successful in meeting this constraint within Math 1. Most of the teachers who are now involved in the project for a second year not only find it easier to prepare for and teach a CPM course than they did in their initial year of teaching Math 1, they also find teaching a CPM course requires less effort than teaching their usual courses.

Third, *the program must be teachable by most teachers.* A program built around a cadre of select teachers may be truly wonderful, but is unlikely to have an impact on students in typical schools in large metropolitan school districts. Our guideline is that 80% of the teachers must be able to adapt to this program if they want to.

Fourth, *the program cannot be too costly.* Just as the amount of time and effort required by teachers are constraints, so is money. Thus, we do not assume that our students will have graphing calculators, nor do we assume that students will have much access to computers. We do assume that students will buy or have available standard scientific calculators. The price of graphing calculators is dropping so rapidly, however, that our assumption that they are too expensive may not be valid in five years.

Finally, *the program must not cause a perceived loss of control.* This constraint is significantly different from the previous ones because here we are dealing with perceptions, particularly those of teachers. We are also concerned with the perceptions of administrators, colleagues, students, and parents. Our program involves students working in groups a significant fraction of the time, which means the classes may sound noisier and appear less orderly than a traditional class based on a lecture/individual practice format. If the perception within the school is that CPM teachers do not have control of their classes, then the program would face enormous difficulties. Two consequences follow. First, everyone must come to a new consensus on how a well-managed classroom looks and sounds; this is largely a matter of reeducating school personnel. Second, teachers must learn the somewhat different skills required to successfully manage a class of independent learners.

Assumptions about learning. While it is possible to design many courses which satisfy the assumptions about the world listed above, our course is also guided by three major assumptions about learning. These are our guiding principles. It should be clear on reading them that these principles are conceptually independent, but that taken together they represent a rather large step in reshaping the teaching of secondary school mathematics.

Guiding principle one: students should be actively involved in their learning. We take this statement very seriously, and in doing so, we find that it has enormous implications. It has an impact on what materials we use, what behaviors we expect from the students, and what teaching strategies the teachers must employ.

A teacher viewing our materials for the first time commented, "After looking at this material, I can't imagine how I can lecture on it," and one of our directors responded, "Well, I can't either." These materials are not designed for a traditional classroom. They are made for a very different kind of learning environment, one in which most learning takes place in small groups and where learning occurs by students working through sets of problems.

Because the learning environment for which our materials are designed differs fundamentally from that of traditional secondary mathematics classes, the teachers who adopt these materials must learn how to teach in a new way. This must be a way that reinforces the message to students that most of the concepts in the course are relatively easy and that they, the students, can learn them *on their own*. This is a difficult message to convey to both teachers and students, because it runs counter to years of experience with traditional mathematics classes. Teachers are certainly not irrelevant in our program; they enter the picture in critical ways and at critical times. But our experience has validated our initial assumption that students can learn most of the traditional algebraic skills without having someone model formal step-by-step procedures. Our anecdotal evidence is that our students know these skills better—a statement we believe to be true and are now studying in some detail.

Guiding principle two: teach fewer topics and teach them better. Much of the secondary mathematics curriculum is a historical anachronism dating from the days before computers and calculators were widely available. Clearly, these are times of transition. Courses that are relevant for the 1990s will evolve into different courses for the future, as technology becomes cheaper and more widespread and as we understand what implications these changes mean for what should be taught. No one today can say for certain what constitutes an appropriate core of knowledge for high school students entering the twenty-first century. So the next 20 years are going to be a wonderfully exciting time in mathematics education as we try to figure out what is important and what is not.

But today there are still a great many topics which remain in textbooks merely to satisfy everyone on district adoption committees. Worse yet, many teachers feel compelled to cover certain topics simply because they are in the chosen text and do so even at the expense of devoting more time to more important topics. When we first met with our teachers to discuss the essential topics to be covered in three years of college preparatory mathematics, we pushed very hard to get them to evaluate the importance of the proposed topics and ruthlessly excluded topics for which there was not overwhelming support. In general, the three directors agreed with the teachers on everything except factoring; there was a near riot when we proposed that it be omitted completely. This debate continues and is an ongoing reminder that curriculum change is an art of the possible.

There are a number of topics in our Math 1 class that are not treated as thoroughly, or perhaps as well, as they could be. In our second revision (third version) of the course, we added some more material to support and strengthen what is already there. This means that we had to move some other material out of Math 1 and into either Math 2 or Math 3.

Guiding principle three: big ideas take a long time to learn. This means that if we want students to learn something important, say the notion of a ratio, we cannot give them a unit on ratios for two weeks and have any realistic expectations that they will understand ratios. Some students may have a fairly good idea, but many students will have no real understanding even if they have managed to learn a pattern imitating how to solve a certain kind of problem. So it will take a very long time, weeks or months, for most students to learn that big important idea. That is, *mastery comes later for most students.*

This concept is very difficult for most mathematics teachers to internalize because of the way they themselves have been taught. Of all the adjustments they make in teaching with the CPM materials, the teachers have reported the most difficulty in adapting to this notion. We struggled with this notion as a group when we produced suggested tests for the teachers to use; we still had many mastery questions for unit six on the unit six test, for example. More appropriately, upon completion of unit six we should be looking for mastery of some of the concepts and skills introduced in units three or four and minimal understandings of the material in unit six.

If we assume that mastery of an important topic will occur after a long period of time, and we still expect to introduce several topics, it follows that the materials must involve an enormous amount of spiralling. In this way, a new idea is reinforced, in either a familiar or a new form, every day for a few weeks, then every other day, and eventually once a week, for the rest of the year. It is a source of both pleasure and surprise to our teachers that this idea actually works, that students eventually acquire understanding of an idea without further formal instruction.

There is nothing magical about using these three principles in combination. We could have chosen any one or two of them, but we chose to adopt all three. Having done so, the structure of our curriculum was essentially forced on us. In combination, these three principles have made for dramatic changes in the form and content of traditional college preparatory math courses. Given the constraint that most teachers should be able to teach the materials, these changes are probably as far as we can go in one generation of teachers.

Changes

The key question, of course, is how can all three of our guiding principles be transformed into a well integrated, three-year course? Despite our strong belief that fewer topics should be taught during these three years, we did add two topics: problem solving and three-dimensional geometry. These two topics were chosen because problem-solving skills and facility with spatial visualization are highly useful in several contexts, both inside mathematics and outside. Both of these are also crucial skills in many college mathematics courses.

Problem solving was added because we believe that problem-solving skills are useful not only in themselves, but also as *learning tools.* People need higher-order thinking skills when they meet new situations, and students are most likely to meet new situations in their studies. Thus they will need to have facility with a variety of problem-solving approaches if they are to be able to learn independently.

Three-dimensional figures seem to have been banished when solid geometry was dropped from the curriculum. Most students in a beginning college calculus class have enormous difficulty visualizing what happens when a curve is rotated about the x-axis. This difficulty occurs largely because spatial visualization is an intellectual skill which they have never practiced. Beginning in Math 1, we ask the students to work on problems which involve relationships in three dimensions. About six weeks of Math 2 are intensively devoted to these same issues, which will be carried into Math 3.

Another significant change is the introduction of some geometry, particularly the Pythagorean theorem and similar triangles, into Math 1. We introduced the Pythagorean theorem in order to allow for the creation of more interesting problems involving quadratics, while familiarity with similar triangles is clearly needed for students to develop a full sense of what ratios are. At the same time, much algebra is reviewed throughout Math 2, as far as possible in the context of the geometry which is being developed.

Finally, we have expanded an emphasis on written explanations to run throughout the entire three-year program. Students begin writing short explanations in Math 1, such as, "In one or two complete sentences, explain to your friend who was absent yesterday how...." They are also asked to compare processes and make connections between certain ideas. In Math 2, students progress to the point that they are able to write full explanations of their reasoning, culminating in formal geometric proofs. The emphasis on students writing about processes, connections, and reasoning, including proofs, will continue throughout Math 3 as well.

Given these changes and additions, together with our view that fewer topics should be taught, it will come as no surprise that several topics traditionally taught in Algebra 1 are either deferred to other years, deemphasized, or eliminated entirely. Factoring is deemphasized as much as we dare. Terminology is cut to the minimum. The whole area of 'simplifying' expressions has been either deferred to Math 3 (for example, rational and radical expressions) or drastically reduced and approached from the problem-solving perspective of looking for subproblems. We try to emphasize that in certain contexts, some ways of writing an expression may be more useful than others, so students should know how to convert algebraic expressions from one form to another.

Implementation of the Project

The ideas discussed above are implemented almost exclusively through student problem sets. Each unit consists of a list of 100 to 150 problems which are carefully sequenced to introduce the necessary concepts, provide some routine practice, force students to reflect and explain, and review and integrate old ideas. It is assumed that the students, by working in groups, will discover and articulate most of the important ideas for themselves by working through these problems. An occasional example is given as a model, but these are quite rare—approximately 20 in the entire year of Math 1.

Cooperative groups have a far larger responsibility in the CPM program than when used in most other courses; it is assumed that the *knowledge will be generated within the group*, not that the group members will be explaining knowledge generated by the teacher to each other. This shift will not occur unless the teacher makes a serious effort to restrain his/her natural tendency to answer questions; rather, the questions need to be refocused and given back to the group to answer. It should be no surprise that this change in classroom atmosphere causes enormous strains during the first month among students who are accustomed to being told which algorithm to follow. An extremely common complaint by students to our teachers during the first month of school is that they are not *real* teachers because they are not teaching (that is, not telling) them anything. It takes determination on the part of the teachers to survive the first two months until the students accept the fact that they will have to learn how to learn in a new way. After that point, the teachers report many students get restless if the teachers try to do any traditional teaching. Once the shift of knowledge generation to the students within the group setting has occurred, it tends to be fairly permanent.

Students need more than problems, of course. Most lessons during the year encourage or require the use of materials other than pencil, paper, and calculator. Problem sets are designed to use the work done with manipulatives to help students extract the important mathematical ideas. The teachers are encouraged to make their classrooms 'manipulative-friendly' and to give real permission for students to use any useful tools they can find for creating understanding.

The unit outlines for the first revision (second version) of Math 1 appears in Appendix 1. Some sample problem sets are included in Appendix 2.

A Typical Day

While there really is no such thing as a typical day, classes which utilize our program well often have several common features. A class period usually begins with some sort of focussing review. This is generally done by having the students consider problems similar to those from a few days or a few units back which may be helpful in getting started on the day's lesson. Other days the teacher will simply pose review problems. Typically, these are extremely routine.

The teacher then gives a very brief introduction to the lesson, rarely more than a minute or two long, trying to put the ideas into a larger context. For example, the teacher may recall the notion that solving equations involves 'undoing' the complications which make an equation hard to solve and discuss general strategies for undoing. Then students break into their groups, usually groups of four but sometimes pairs, to begin working on the problems in the text, *always* with the aid of their calculators and often with the help of readily available manipulatives.

Daily outlines are suggested in the teacher notes that accompany the texts. A typical assignment consists of about ten problems, the first five of which are more difficult and/or introduce new material, so are done in cooperative groups in class. The remaining problems review old ideas, ranging from those of the previous day to ideas which were introduced months before and can be done as 'homework'. During the class period, the teacher circulates around the room, listening to groups and helping groups get 'unstuck' if they are having difficulties. If the teacher observes that several groups are having the same difficulty, he/she generally calls the class back together to point out where the particular difficulty lies and have the whole class discuss strategies for dealing with it. On rare occasions, if the frustration level is very high, the teacher will simply walk through the problem for the class. At the end of the period, the teacher calls the students back together, collects any materials which need to be put away, and sometimes summarizes the day's work.

In general, our teachers report the need to do less work to prepare for class. However, during class they are often asked to respond to unexpected questions, or they might need to refocus discussions on the spur of the moment, or give reasonable hints on the spot. Most CPM teachers also report a need to be more flexible in their attitude toward goals reached on a given day. This is important because virtually all have had the experience of a vigorous debate erupting on a particular issue in one class while another class completely ignores the same topic.

Now and the Future

What are the results to date of 4000 students using CPM Math 1 materials for their algebra courses and 1500 students using Math 2 materials for their geometry classes? First, the existence of the materials seems to have tapped a deeply felt dissatisfaction with current texts. Over 300 inquiries about the project have been made from around the country and approximately 100 teachers from 30+ school districts are interested enough to apply to teach the courses in the 1991–92 year, a commitment which involves three days of training in August, plus five days of release time during the school year, plus paying for the materials. Given the extraordinary cuts in California school funding in 1990–91, the response is quite amazing.

While this interest is very gratifying, it is somewhat premature. At the moment, the expansion districts are taking much on faith. We still have not answered some fundamental questions about the project and are currently studying both students and teachers to get these answers. For example, we do not even know at the moment how well CPM students learn the concepts which we consider most important, let alone whether they know these concepts better than students coming from more traditional classes. We need to answer this question and several others before we consider a truly widespread expansion of the project. The next few years of the project, therefore, must involve a very serious look at what happens to CPM students during their year(s) learning from our materials.

The anecdotal evidence is quite good to date. So far, most of the teachers are quite enthusiastic. They assert that their students are understanding better even though scores on some standardized tests have dropped. Most teachers report that weak students are persisting longer in their classes and, if not doing well, are not failing either. Considering the amount of group work involved, the courses seem to have been particularly successful with English as a Second Language students. Hispanic students are also apparently doing much better than in traditional classes.

For now, we are taking these claims with a grain of salt because all of these reports come from people who have a vested interest in making the project succeed. We will not feel truly comfortable about these assertions until a few more years have passed and we are able to study the impact which occurs when less committed and less able teachers have also used the materials.

One result is already clear. We are asking such a radical change of teaching style that very few teachers can do it alone. A few have tried to use the materials on their own without support from other teachers in the school and without contacting others who are also working with CPM materials. They have all given up within a few months. So reasonable support is necessary for most teachers to make the transition.

In conclusion, the CPM program is a three-year course of college preparatory mathematics designed to replace the traditional sequence of college preparatory math courses. Its goal is to better meet the needs of all students, particularly those from groups who have been underrepresented in mathematics in the past, in a highly technological society. It is based on two types of assumptions about curriculum change, which we have identified as constraints and guiding principles. We believe that in order for a new curriculum to be successful, it must meet certain constraints imposed by schools and society: new courses must be compatible with existing courses; they must be teachable by a large fraction of the teachers currently in the schools; they cannot impose excessive time and cost burdens; and they cannot cause a perception of loss of control by the teachers. In addition, our work is guided by three principles about learning: that students should be actively involved in their learning; that we should teach fewer topics and teach them better; and that mastery of ideas takes time.

Our assumptions about learning naturally led to great changes in the form and content of traditional mathematics courses. Considering the constraints imposed on curriculum change, we believe the changes we have incorporated in our materials to be as much as can be successfully accomplished in the near future. Major content changes are the use of algebraic and geometric concepts to support and extend each other and the emphasis on students writing explanations. Pedagogical changes include the use of problem solving as a learning tool, the use of the cooperative groups as the setting in which students generate knowledge, and the spiralling of concepts throughout the three-year course, so that topics are continually reinforced and extended.

Outcomes to date have generally been positive, although our evidence at this point is mostly anecdotal. We are in the process of seriously evaluating the program and have applied for further research support to accomplish this. At this time, after four years in the project, we believe we are on the right track and that our efforts will pay off in successfully changing the way the college preparatory mathematics curriculum is perceived and taught by a large fraction of secondary math teachers.

For further information about the CPM program, write to:

College Preparatory Mathematics Project
CRESS Center
University of California
Davis, CA 95616

APPENDIX I

Algebra 1
(College Preparatory Math 1)

Table of Contents

Timeline

Foreword to the Teacher

Introduction and Overview

Materials List

Unit 1 **Openings: Data Organization** Problems OP-1 through 45
This introductory unit shows students that this course is different from a traditional Algebra 1 course. It teaches them to work in groups and pairs as they collect, organize, and graph data (bar graphs or scatter plots), solve some logic problems and develop calculator skills, including order of operations and use of exponents.
(2 weeks)

Unit 2 **The Burning Candle: Patterns and Functions** Problems BC-1 through 91
Students explore from pattern to rule to function to graph and vice-versa, making use of their scientific calculators. The Burning Candle Problem uses the pattern formed on a graph to make a prediction.
(3 weeks)

Unit 3 **Choosing a Phone Plan: Writing Equations** Problems CP-1 through 128
This is a very important unit. Building on their patterning skills, students learn how to solve standard word problems by Guess and Check. Then they use the technique to write equations for the problems. Students also learn more standard ways of solving equations, beginning with the concrete approach of "cups and tiles" and followed by working backwards in the "cover-up" method. Students write equations to solve the problem of which phone plan to choose.
(4 weeks)

Unit 4 **At the Movies: Ratios and Similarity** Problems AM-1 through 81
The concepts of ratio and similarity are crucial for anyone going on in mathematics or the sciences. Students examine numeric and geometric ratios in a variety of activities. They explore the concept of similarity by enlarging and reducing simple figures on dot paper, then compare ratios of perimeters and areas of similar figures. They examine similar right triangles in preparation for an introduction to the concept of slope later in the year.
(2 weeks)

Unit 5 **Tiling Rectangles: Factoring Quadratics** Problems TR-1 through 71
Here students see the concept of multiplication and factorization of quadratic polynomials in terms of finding the areas of rectangles whose side lengths are variables. To build a geometric understanding, we start with tiles and diagrams and make connections to the algebraic representations.
(3 weeks)

Unit 6 **The Lunch Bunch: Graphing and Estimation** Problems LB-1 through 81
This is another very important unit. We return to the question of graphing more complicated equations. Students see interrelations between the concepts of graphing and solving equations with one or two variables. They also use their calculators to solve equations approximately when they cannot do so directly. The introduction to the unit is a short play, The Lunch Bunch, that introduces a problem that can be solved by considering systems of equations. (We don't solve the Lunch Bunch Problem completely until the final unit of the course, after we've gathered all the necessary tools.)
(3 weeks)

Unit 7 **A Picture is Worth a Thousand Equations: From Diagrams to Equations** Problems PT-1 through 107
In Unit 3, students learned how to translate verbal descriptions to equations. Here the emphasis becomes translating verbal descriptions to diagrams and then to equations. We use the Pythagorean Theorem as a vehicle for introducing quadratic equations and developing the need for radicals.
(4 weeks)

Unit 8 **The Grazing Goat: Area and Subproblems** Problems GG-1 through 102
We look more closely at the concept of area and use it to introduce the concept of subproblems in a visual way. Students then use the idea of subproblems to simplify many problems, including rational expressions. The Grazing Goat Problem is a good example of a problem that includes several subproblems.
(3 weeks)

Unit 9 **The Return of the Burning Candle: More Ratios and Slope** Problems RB-1 through 116
Students use similar right triangles to develop the notion of the slope of a line and to write equations, this time from a graph. We return to the Burning Candle Problem to develop an algebraic solution through writing and solving equations.
(3 weeks)

Unit 10 **The Election Poster: More About Quadratics** Problems EP-1 through 62
Students have been solving quadratic equations by Guess and Check, by graphing, and by factoring. In this unit students are given the quadratic formula as a fact so they can extend their equation solving skills. The problems students encounter, such as the Election Poster Problem, tie together work the students have been doing throughout the course. Two activities extend students' understanding of the graphs of quadratic equations, and further connections are made as students work through a series of subproblems to develop a proof of the quadratic formula.
(2 to 3 weeks)

APPENDIX II

For problems BC-72 and BC-73, write an algebraic rule if the problem is stated in words and write a verbal statement for a rule given in symbols. Then make a table for x-values between -4 and 4 (inclusively). Be sure to include some fractions and decimals. Graph the results.

BC-72. $y = -2x^2 - 3$

BC-73. $y = -5x + 9$

BC-74. If a ball is tossed into the air at 80 feet per second, its height "y" above the ground after "x" seconds is given by $y = 80x - 16x^2$. (This is a formula used in physics. In Algebra II, we'll see how it was derived.)

a) Make a table for $0 \leq x \leq 5$. Be sure to include some non-integer values such as 0.25, 0.5, 4.8, etc.

b) Graph your information from a). Scale the x-axis 5 squares of graph paper per unit and the y-axis 5 units for each square of graph paper.

c) How high is the ball after one second?

d) When is the ball 96 feet high?

e) When is the ball 200 feet high?

f) Describe the shape of the graph and what it represents in terms of tossing a ball into the air.

g) When does the ball reach its maximum height? What is this height?

h) For what values of x is $y = 0$? What does this represent about the ball?

i) If $x = 6$, what is the value of y? What might this represent?

Symbol: ". . ."

The symbol ". . ." is called an ellipsis and indicates that values have been omitted. Missing values follow the same pattern as those shown.

Example: {3, 3.5, 4, . . . , 8.5, 9} means that 4.5, 5, 5.5, 6, 6.5, 7, 7.5, and 8 have been omitted.

BC-75. For each equation, make a table and graph it using the input values provided.

a) $y = x^2 + 2$ for $x = -4, -3, -2, -1.5, -1, -0.5, 0, 0.5, 1, 1.5, 2, 3, 4$.
b) $y = 3x^2 - 5$ for $x = -2, -1.5, -1, -0.5, 0, 0.5, 1, 1.5, 2$.
c) $y = 2x^2 - 3x - 4$ for $x = -2.0, -1.5, -1, \ldots, 2.5, 3$.
d) $y = -x^2 + 2x - 3$ for $x = -4, -3, -2, \ldots, 2, 3$.

BC-76. Use your calculator's y^x or x^y key to find:

a) 2^3
b) 3^2
c) 3^0
d) 0.5^3
e) 2^{-1}
f) 0.5^0
g) $(\sqrt{3})^4$
h) $1.5^{3.2}$

BC-77. Make a table of values for $x = -5.0, -2.5, -2.25, -2.0, \ldots, 2.5, 2.75, 3.0$ to create a huge graph of $y = 2^x$. If you are using standard graph paper, scale the axis so that four squares equal one unit. Use the graph to answer the following:

a) What happens to y as x gets larger? As x gets smaller?

b) Where does the graph cross the x-axis?

c) Use the graph to estimate the x-value when the y-value is:
1) 5
2) 2
3) 1/2
4) 3.5

d) Use the graph to find A and B so that:
1) $2^A = 7$
2) $2^B = 12$

BC-78. Make a table of values for $x = 0, 0.5, 1, 2, 3, 4, 5, 6, 7, 8, 9, 9.5, 10, 25, 32$ and create a large graph of $y = \sqrt{x}$. Scale the graph as you did in problem BC-77.

a) Use the graph to estimate the values of x so that:
1) $\sqrt{x} = 1.7$
2) $\sqrt{x} = 2.2$
3) $\sqrt{x} = 0.5$

Can you find more accurate solutions using a calculator? If so, what are they?

b) Try the values $x = -1$ and $x = -3$ for the function using your calculator. Explain your results. Why are there no points on this graph to the left of the y-axis?

AM-18. Note that the size of the dot grid in AM-17 is different from that used in problem AM-11.

a) In AM-11 and AM-17 you did the same task--making a new figure half again the original --but with different size grids. Compare the ratios you got in AM-11 and AM-17.

b) In AM-11 and AM-17, you made a new figure $\frac{1}{2}$ the original. Suppose you did it again on a grid where the dots are one foot apart, what do you think the ratios you compute for perimeter and area would be?

AM-19. a) Write one or two sentences explaining what you think will happen on <u>any</u> size grid to the ratios $\frac{\text{Perimeter of new figure}}{\text{Perimeter of original figure}}$ and $\frac{\text{Area of new figure}}{\text{Area of original figure}}$.

b) Answer the same question if the new figure has side lengths 2 times the original.

c) Answer the same question if the new figure has side lengths 3 times the original.

d) Answer the same question if the new figure has side lengths 10 times the original.

AM-20. Make a conjecture what the ratios $\frac{P_{new}}{P_{original}}$ and $\frac{A_{new}}{A_{original}}$ will be if the new figure is N times the original.

AM-21. Solve these Diamond problems:

a)

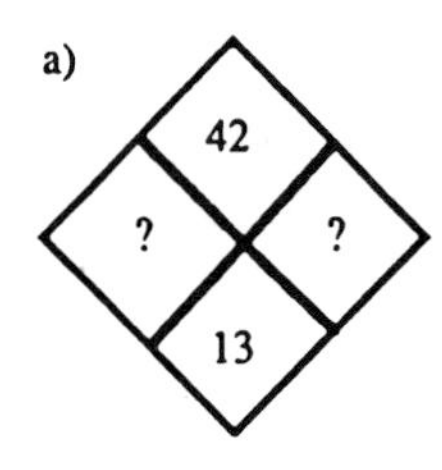

b)

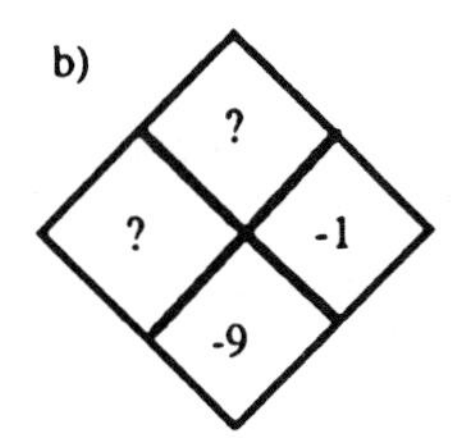

AM22. Find the following products:

a) 6(x + 2)

b) 5(a + 3)

c) x(3 + 5)

d) y(6 + 4)

j) Cut (or tear) ∠A" off Δ A"B"C". Tape or glue it on a third piece of graph paper so that one side is on the y-axis and the vertex is below the height of ∠A. Use your straightedge to extend the other side of ∠A" until it intersects the x-axis. Cut or tear ∠C" off Δ A"B"C" and place it into the angle formed by the x-axis the extended side of ∠A". If ∠C" doesn't fit, go back and make sure the sides of ∠A" are placed precisely. The resulting triangle is your new Δ A"B"C".

k) Which, if any of your three triangles, Δ ABC, Δ A'B'C', and Δ A"B"C", are similar? Explain your answer.

l) Find: |AB| = |A'B'| = |A"B"| =
|BC| = |B'C'| = |B"C"| =

m) Compare the ratios $\frac{|AB|}{|A'B'|}$ and $\frac{|BC|}{|B'C'|}$.

Now compare the ratios $\frac{|AB|}{|A''B''|}$ and $\frac{|BC|}{|B''C''|}$.

Finally compare the ratios $\frac{|A'B'|}{|A''B''|}$ and $\frac{|B'C'|}{|B''C''|}$.

AM-32. Suppose you are given two similar triangles: Δabc and ΔABC with ∠a = ∠A, ∠b = ∠B, and ∠c = ∠C.

a) Look back at exercises AM-25 and AM-31 to make a conjecture about what is probably true.

b) With the help of the other members of your group, write <u>one</u> conjecture for your whole group as carefully as possible.

AM-33. Find |TC| .

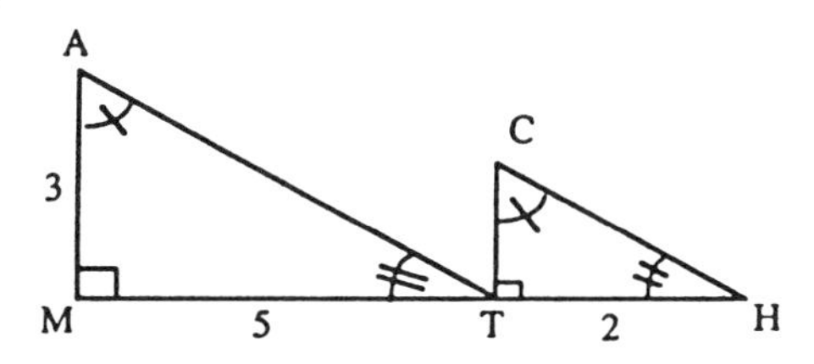

c) Let x represent the number of rides and y represent the total amount spent (in dollars). Write an equation for each admission option. Use the patterns from (a) and (b) to help you.

d) Graph both admissions options, using (a) and (b) or the equations in (c). Scale the x-axis 1 unit per ride up to 25 rides, and the y-axis 2 units per dollar up to $20.00.

e) Use the graph to find the number of rides that cost the same regardless of the option Jim chooses.

f) Use your equations to solve for the answer to (e). Refer to LB-44 (f) through (i) if you need help.

g) Now conclude the problem by writing two or three sentences to advise Jim as to what he should do. Several responses are possible.

LB-51. GOING SHOPPING

Barb and Betsy went shopping. Barb went to Tower Records and bought some tapes. Afterwards she spent $12 on a sweatshirt. Betsy bought some tapes and she also bought a pair of earrings for $3. We don't know the number of tapes each one bought, the price of each tape, or the total amount each one spent. Let's assume the tapes all cost the same amount.

a) Let x represent the price of a tape (in dollars) and let y represent the total amount of money each one spent (also in dollars).

(1) Suppose Barb bought 3 tapes. Write an equation relating x and y to show the total amount of money she spent.

(2) Suppose Betsy bought 5 tapes. Write another equation relating x and y to show the total amount of money Betsy spent.

b) Graph both of the equations from (a) on the same set of coordinate axes. Scale the x-axis 4 units per dollar out to $8.00 and scale the y-axis 1 unit per dollar up to $40. Extend the lines beyond their point of intersection.

c) Find the coordinates of the point of intersection. What is the real-world meaning of the point of intersection? Reply in a few complete sentences.

d) Suppose Barb and Betsy each spent the same total amount of money. This means that both y-values are the same. Now you can figure out how much a single tape costs. Do so in two ways:

(1) estimate from the graph, and

(2) write one equation in which you let the expressions involving x equal each other and solve for x.

LB-74. Without drawing a graph, determine which of the points below are on the graph of the linear equation $y = -2x + 3$.

(0,3), (3,0), (-3,3), (1,1), (5,-7), and (5,7).

LB-75. a) What are the two most important ideas you have learned in this unit?
b) How do these ideas relate to other important ideas you have learned this year in algebra?

LB-76. MORE OR LESS

Judy has $20 and is saving at the rate of $6 per week. Jeanne has $150 and is spending at the rate of $4 per week. After how many weeks will each have the same amount of money?

Solve this problem in a step by step manner by writing an equation for each person and drawing the graphs to estimate the solution. Scale the x-axis 2 units per week up to 20 weeks and the y-axis 5 units per $25 up to $160. Now show a method for finding the solution exactly using algebra.

LB-77. LESS OR MORE

Judy has $20 and is saving at the rate of $6 per week. Jeanne has $150 and is saving at the rate of $4 per week. After how many weeks will each have the same amount of money?

Solve this problem in a step by step manner similar to the preceding problem. Scale the x-axis 2 units per week up to 20 weeks and the y-axis 5 units per $25 up to $160.

LB-78. Carol and Jan leave from the same place and travel on the same road. Carol walks at 2 miles per hour. Carol left 5 hours earlier than Jan but Jan bikes at 6 miles per hour. Suppose we want to use algebra to find out how long it will take Jan to catch up with Carol.

a) Because we are told that Jan catches up with Carol, we know they have traveled the same distance up to that time. Therefore, the diagram shows the same length arrows. Start your solution by writing, "Let x represent the time (in hours) for Jan to catch up with Carol" on your paper. Then copy the diagram on your paper.

Carol

Jan

b) Write the length of time and the rate they are traveling on your diagram for each of them. Be sure to write units each time you write a number (don't just write "5", write "5 hrs").

c) Write an expression using x to represent the distance Carol travels. On your paper, write "Carol's distance = _______ " so that you know what the expression represents.

d) Write an expression using x to represent the distance Jan travels. Be sure to write "Jan's distance = _________."

e) Write an equation that says they are traveling the same distance and solve the equation.

Belmont University Library

GG-84. The length of the diagonal on the classroom floor is 30 feet. The height of the room is 10 feet. A fly is on one corner (B) and flies to the opposite corner (A). How far is it from A to B? Would it matter which corner the fly starts at if it flies to the opposite corner?

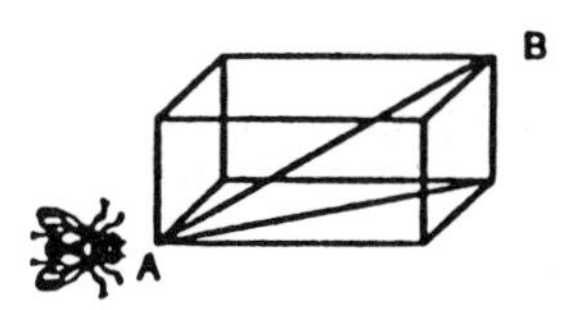

GG-85. Solve the following systems of equations.

a) $3x - 2y = 4$
$4x + 2y = 10$

b) $p + q = 4$
$-p + q = 7$

c) $y = x + 2$
$x + y = -4$

<u>Example</u>: Suppose we want to solve this system of equations:

$$3x + 2y = 14$$
$$2x + 3y = 11$$

Again, the subproblem is to ELIMINATE y in one equation. But now to do this, we need to do something to BOTH equations.

Multiply first equation by 3 (why 3?) to get: $9x + 6y = 42$
Multiply second equation by -2, to get: $-4x - 6y = -22$

How can we now get rid of y terms?
Answer: By <u>adding</u> the two new equations.
We get: $5x = 20$

Then, we know $x = 4$ and can easily find that $y = 1$.

GG-86. In your group, discuss the example above. Would it have mattered if we chose to multiply the first equation by 2 and the second equation by -3? Try it! (Write it out to check.)

GG-91. Solve for x and y.

a) $3x + 3y = 15$
$x - y = 6$

b) $2x^2 - y = 5$
$x + 2y = -5$

GG-92. A square with a side of length "x" cm on a side has a circle of diameter x cm cut out of it. Find the fraction of the square left if:

a) $x = 2$

b) $x = 20$

c) $x = 10$

d) For any x (in terms of x)

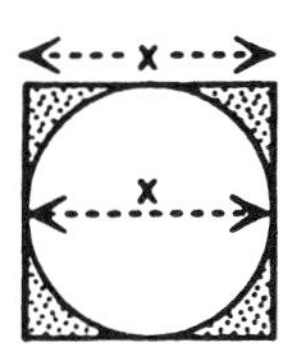

GG-93. A cookie baker has an automatic mixer that turns out a sheet of dough 12 inches wide and 3/8 inch thick. Usually his cookie cutter cuts 3 inch diameter circular cookies as shown at the right. His boss complained that too large a fraction of the dough had to be re-rolled and ordered him to change to making 4 inch diameter cookies.

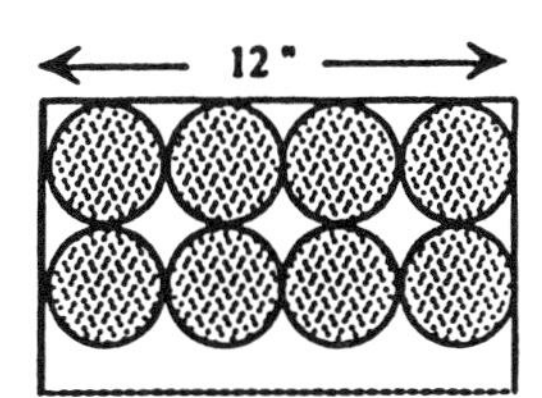

a) What fraction of the dough had to be re-rolled for the 4-inch diameter cookies? 3-inch diameter?

b) Answer the same question for 6-inch, 2-inch and 1-inch diameter cookies.

c) Compare your answers.

d) Is his boss correct? Explain your answer in one or two sentences.

References

1. *Curriculum and evaluation standards for school mathematics*, National Council of Teachers of Math., Reston, VA, March 1989.

2. *Everybody counts: A report for the nation on the future of mathematics education*, Board on Mathematical Sciences and Mathematical Sciences Education, National Research Council, National Academy of Sciences, Washington, D. C. 1989.

3. Bill Honig (Superintendent of Public Instruction), *Mathematics framework for California public schools, kindergarten through grade twelve*, California State Department of Education, Sacramento, CA, 1992..

Department of Mathematics, California State University, Northridge, CA 91130
E-mail address: kasimatis@ucd.math.edu

Department of Mathematics, University of California, Davis, CA 95616
E-mail address: sallee@ucd.math.edu

CBMS Issues in Mathematics Education
Volume 3, 1993

Small Groups for General Student Audiences 1

RICHARD J. MAHER

ABSTRACT. This article begins by noting the origins and evolving use of small groups in the quiz sections of large lecture classes of lower division mathematical sciences courses at Loyola University, Chicago. Data is analyzed and the results, some unexpected, are discussed. Comments also are offered on current and planned activity. The entire presentation occurs in a context indicating the general applicability of the methods presented here: methods that require minimal extra spending.

The use of small groups in mathematics courses, an area of interest for some time [**3** and **5**], has received even more attention over the last few years [**1**, **2**, **4**, **10**, **11** and **12**]. Small groups often are used in connection with large lecture sections of lower division courses, particularly when educationally disadvantaged students are involved. Their general purpose is to develop those academic and interpersonal characteristics that commonly are associated with students who do well in mathematics. Increased resources usually are devoted to these efforts, and the end result often is an improved performance by the students in the small groups.

Based upon my experiences many years earlier while a graduate student, I decided during the summer of 1986 to use small groups in a class that I was teaching. The results were favorable, and over the past seven summers I have used small groups in various ways in precalculus, calculus, multivariable calculus, calculus-based statistics, and linear algebra courses. (Class sizes ranged from 11 to 31; no extra hours or support were available; for details, see [**9**].) The continued positive results led me to wonder what impact small groups might have on lower division courses taught during the regular academic year, particularly large lecture sections. The Department Chair and the Dean both agreed to my request to study their possible impact in a large lecture section that I would teach during the spring term of the 1990–91 academic year and to use the information obtained as a basis for their future use.

Large lecture sections at Loyola University, Chicago are three-credit courses in elementary statistics, precalculus, and nonmajors calculus that meet for either two 75-minute periods or three 50-minute periods each week. Each

©1993 American Mathematical Society
1047-398X/93 $1.00 + $.25 per page

student also enrolls in one of five quiz sections that meet once a week. Maximum class size is 100 and maximum quiz section size is 20. Quiz sections are met by a teaching assistant, with one TA assigned to each large lecture section. The idea I had was to see if some of the learning problems we found associated with large lecture sections might be alleviated by using small groups in the quiz sections of these courses. Since this approach would significantly increase the responsibilities of the TA, an undergraduate major was hired to help in the quiz sections.

The large lecture section I taught during the spring 1990–91 term, to which I had been assigned nearly a year earlier, was a second-term course in nonmajors calculus. The prerequisite for this class was the completion of a first-semester calculus course with a grade of D or higher. The 90+ students in this class were not specially chosen, nor did they know that the quiz sections were to be run in a somewhat different fashion. The five quiz sections, which consisted of 15 to 20 students, were divided into small groups of three to five students at the first session. Throughout the term, the students worked together in their groups on assigned problem sets. These sets were designed to be both cyclic and, as much as time permitted, cumulative. The TA and the undergraduate assistant circulated among the groups, offering encouragement and appropriate comments. They urged the students to work together while at the same time stressing the importance of an individual understanding of the material. They also answered questions and, when appropriate, asked them. The students were graded each week, both individually and collectively, on the work done in the quiz sections; this grade constituted some 12% of their total grade.

The results from this trial section were encouraging. There were no complaints on departmental teaching evaluations about what essentially were required quiz sections (a new experience), which meant the students felt the quiz sections served a purpose. Both the TA and the undergraduate assistant reported that they were kept busy during most sessions from the second week on. The TA, who had assisted in nonmajors calculus the two previous semesters, and I both noted increased interest in the material and better formulated questions. The in-class examinations in general were much better written. The number of withdrawals and failures was lower than those I had experienced previously in large lecture courses, while the GPA was higher. One outcome was striking: the percentage of students turning in nonrequired homework (daily assignments independent of the quiz sections) rose from a typical 20%-35% range to nearly 70%. In other words, the pattern of increased student interest, good feelings, and positive results that occurred at other institutions using small groups in various settings also occurred in this class. These results were obtained without extra class or quiz section meetings; the only extra expense was the salary of the undergraduate assistant, an amount equal to that paid a new part-time teacher in a lower division course.

Given successful results in this class, the original plan called for using the same approach during fall 1991–92 in my large lecture section of nonmajors calculus and in a large lecture class in basic statistics. General comments based on anecdotal data, numerical data, and information from the other large lecture sections would be made. A decision then would be reached as to whether to use this approach in other sections over the next few terms and to obtain data allowing more specific comments to be made.

This planned approach was abruptly changed. Most faculty were aware of my summer work and followed its academic year implementation with interest. The results convinced the fall instructors in the other two large lecture sections of nonmajors calculus and in the precalculus course that their students might benefit from a similar approach. The Department Chair agreed and we approached the Dean with a request to fund five, as opposed to two, undergraduate assistants for large lecture sections. The Dean agreed and, as a result, five large lecture sections used a small study group approach in their quiz sections during fall 1991–92.

The results of our fall experience were extremely positive and seemed to mirror the results of the experimental section. Faculty, teaching assistants, and student helpers were all pleased with the outcome and with the general atmosphere in the courses. As to numbers, when compared with previous large lecture sections, total withdrawals and failures decreased while grade point averages increased, with both changes significant at the 95% level. The D level grades also decreased, though not significant at the 95% level. Finally, the performance of students in large lecture sections using small groups in quiz sections was similar to the performance of all students enrolled in lower division mathematics courses over the fall 1986–87 through spring 1990–91 period.

Data relevant to these assertions is contained in Table 1, but when looking at this data, it is important to remember that no attempt is being made to present large lecture sections as a desired method of teaching. Large lecture sections are not desirable but often may be a necessary evil, a reality that cannot be ignored. In such cases, we must try to make these sections a quality learning experience. The data given in Table 1 indicates that using small groups in quiz sections of large lecture sections can significantly improve student performance in such courses. That the results are close to those obtained in all lower division courses is a bonus.

It is important to note that small groups were used only for the quiz section population of the large lecture sections in precalculus, nonmajors calculus, and elementary statistics; they were not used in the lectures themselves. Additional meeting hours were not used and extra funding involved only the salaries paid to the undergraduate assistants. The positive results obtained were both immediate and independent of classroom techniques used by individual instructors.

The data accumulated also generated some unexpected information. The GPA for female students in large lecture sections using small groups was

TABLE 1.

	All classes*	Large lectures no small groups	Large lectures small groups
Enrollment	19172	4053	520
W or F grades	5892	1320	125
Fraction of enrollment	0.3073	0.3257	0.2404
D level grades	1977	480	57
Fraction of enrollment	0.1031	0.1184	0.1096
Grade point average	2.50	2.29	2.45

*Data from fall 1986–87 (when large lectures were introduced) through spring 1990–91 for lower division courses.

higher than that for females in various other groupings. These groupings include lower division courses during the period in question, large lecture sections not using small groups, and standard size courses. However, there are several qualifications (see Table 2).

In spite of the qualifications noted in Table 2, these results still raise interesting questions about the impact of small groups on female students. The literature comments (for example [7 and 8]) on a lack of female participation in classroom discussions and on possible favoritism by instructors toward male students. In principle, small groups seem to address both these problems since, by their nature, they not only foster discussions involving all the students but also insure adequate explanations and responses to questions posed.

The comments made in this article are by no means intended to serve as a general statistical study. But they do offer indication that small groups can be a useful method to try and improve student performance in lower division mathematics courses. Small groups offer the sense of interest and personal touch that now seems less and less present in all too many classes. They may also be effective in encouraging female students in these courses. In any event, based upon the results obtained thus far, our department will continue to use small groups in the quiz sections of large lecture sections for the next two or three years. Data generated will be analyzed and a report issued discussing the results. (Note: Initial data for spring 1991–92 indicates continuing positive results.)

There is nothing really special about the methods used by our department or about the students involved; the methods can be applied to quiz sections in any large lecture setting. In principle, there is no reason why small groups cannot be used in a variety of courses at all levels of the curriculum. These approaches need not have a rigid model that must be followed at each im-

TABLE 2.

Average GPA for female students in large lecture sections using small groups:	2.577
Average GPA for female students in standard size lower division courses:	2.505
Average GPA for female students in all lower division courses:	2.489
Average GPA for female students in large lecture sections not using small groups:	2.474

Qualifications:

1. This information is drawn from a subset of the data in Table 1. Information in the Loyola database is kept by student record and by class summaries; data relating grades and sex of the student is not included.
2. The data generating these figures came from grade sheets obtained from individual instructors. For a variety of reasons, lists were not available for all courses from past year. (For example, the Loyola database does not store grade sheets, and some instructors are no longer at Loyola. From grade sheets that were available, some names had to be eliminated, since the sex of the person in question could not be determined.)
3. This information did not come from a designed experiment. Rather, it arose after the fact from the data collected to make judgments on the impact of small groups on student learning.

plementation. For example, comments in **[6]** indicate that homework that is collected, graded, and returned can have a positive impact on student performance. One of our large lecture instructors is using this approach in conjunction with small groups during spring 1991–92; preliminary results are positive. Other well-motivated experiments will, and should, continue. In general, the use of small groups in quiz sections for large lecture classes has generated a good deal of thought about teaching in general.

The preceding comments are not meant to indicate that changes in course approaches should be undertaken lightly. In particular, using small groups in the quiz sections of large lecture classes requires some preparation. For example, both the TAs and the undergraduate assistants should understand the ideas behind small groups; the needs, attitudes, and personalities they will encounter; the effort and responsibility involved; the need to encourage both individual and group development; and the importance of maintaining contact with each student. Course instructors must understand all of the above and also realize that some time and effort on their part is required to make certain that everything works. Some information can be written down; we have a handout for instructors, TAs, and undergraduate assistants that we revise each term. But there is also a need for a good deal of oral communication among everybody involved. It is critical that all parties feel reasonably comfortable talking with one another. There should be one or two formal meetings each term but there is more need for informal discussion and exchange of ideas.

Small group approaches are attractive, since they can help solve some learning problems in large lectures, as well as in many other mathematics classes. But in no way are they the solution to all the problems we face. Other approaches have been suggested that might also have an immediate impact on learning problems. But no matter what approach is chosen, the experiences of others should be utilized to help decide its merits. (In our case, we decided that a small group approach had a good change to succeed.) If a favorable decision is reached about a particular method, that method should be implemented with minimal but well-directed testing to insure suitability; a lengthy formal testing period may only reinvent the wheel. We must, of necessity, use such approaches until the long-term impact of mathematics reform at all levels of the curriculum, if successful, begins to take effect. We simply cannot ignore the needs of the students we now serve.

References

1. R. Asera, *The math workshop: A description*, Issues in Mathematics Education, vol. 1, CBMS, Amer. Math. Soc., Providence, RI, 1990.

2. R. Asera, *Professional development and a study in adaptation*, UME Trends, **2**, no. 4 (1990), p. 1.

3. N. Davidson, *The small group-discovery method as applied to calculus instruction*, American Mathematical Monthly, August–September (1971).

4. F. Gass, *Working in small groups*, UME Trends, **1**, no. 5 (1989), p. 2.

5. J. Gersting and J. Kuczkowski, *Why and how to use small groups in the mathematics classroom*, College Mathematics Journal, November (1977).

6. H. Gore and G. Gilmer, *Effective strategies for teaching calculus at the college level: A Survey Report*; Morehouse College and the Exxon Education Foundation, 1990.

7. R. Hall, et. al., *The classroom climate: A chilly one for women*, Project on the Status and Education of WOMEN, Association of American Colleges, Washington, D.C., February, 1982.

8. P. Kenschaft, *What can I do*, FOCUS, June, (1989).

9. R. Maher, *Small study groups in summer courses*, UME Trends, **3**, no. 3, (1991).

10. S. Mathews, *Group problem solving in the college mathematics classroom*, PRIMUS, Vol. 4, no 1, December (1991).

11. J. Morrel, *Innovation need not be expensive*, UME Trends, **1**, no. 4 (1989).

12. Philip Uri Treisman, *A study of mathematics performance of Black students at the University of California, Berkeley*, Issues in Mathematics Education I, Vol. 1, CBMS, Amer. Math. Soc., Providence, RI, 1990.

Department of Mathematical Sciences, Loyola University, Chicago, Illinois 60626

Ideas, Issues, and Reactions

CBMS Issues in Mathematics Education
Volume 3, 1993

Student Work and Study Habits at a Comprehensive University: A Preliminary Report

HARRIET EDWARDS

INTRODUCTION

This paper describes a survey in progress at California State University, Fullerton (CSUF), along with some preliminary results. The survey was developed to answer questions arising from the CSUF Mathematics Department's experimentation with various reforms in the teaching of calculus. The preliminary results provided intriguing enough glimpses into student demographics and study patterns so that the survey was repeated and expanded in the following academic year (1990-91).

My conclusion from the preliminary work that has been done is that information about student demographics, study habits, and lifestyles can be of crucial importance to reform efforts, and that it is vitally necessary that the effort be made to collect this information. The precise results may vary from one institution to another, but these results will contain valuable clues to how reform efforts might be directed.

HISTORY

By the fall of 1987, the mathematics department at CSUF had begun to notice major changes in enrollment patterns in the freshman calculus course. Enrollments were dropping drastically, causing great concern for the future number of mathematics, science, and engineering majors. Of those students who did enroll, a very high percentage was not completing the course

This work was supported in part by an Academic Program Improvement Grant of the Trustees of the California State University.

©1991 Trustees of the California State University. Permission is hereby granted to copy and use this document for non-profit educational purposes. Inquiries about other uses may be addressed to: CSU Trustees, Academic Program Improvement, 400 Golden Shore, Long Beach, CA 90802-4275.

©1993 American Mathematical Society
1047-398X/93 $1.00 + $.25 per page

sequence successfully, i.e., finishing with grades of C or better. In fact, 43% of students in precalculus, 36% in first-semester calculus, and 43% in second-semester calculus were unsuccessful. Paralleling a situation found across the United States, our calculus sequence had become a major barrier. A reform effort was clearly needed, so the author, newly graduated from the University of California at Berkeley, was brought in to adapt the highly successful UC Berkeley Professional Development Program [**1**, **2**]. The core of the PDP is the workshop program supporting freshman mathematics classes. During the workshops, which meet twice a week for a total of four hours, students work collegially on challenging problems which not only extend and deepen knowledge of course material, but encourage discussion and social interaction as part of the process of solution. Extensive student recruitment efforts were made in coordination with the campus Minority Engineering Program, and word went out to all students that a workshop program was available, but participation levels were disappointing. Workshop sizes did not increase until a room could be made available in the early afternoon. Late afternoon workshops did not draw students because they wanted to avoid rush hour on the freeways! Others had jobs at those times which prevented them from participating. It quickly became apparent that information on student demographics and work and study habits would be necessary for the effective implementation of this or any other reform program.

At about the same time, the mathematics department was finding that it needed to know more about its students and their enrollment patterns. Why were enrollments dropping in freshman calculus but staying steady in sophomore courses? Would this trend be likely to continue? How many students were arriving with experience in AP mathematics courses and how was it affecting their university work? How much high school mathematics had they taken and how long ago had they taken it? How many 'nontraditional' students were taking calculus and how could they best be served? The department's need for such information dovetailed nicely with Edwards' interests: she in fact had already been conducting informal surveys of her own students, inquiring into work and study habits. From this informal basis, the first pilot survey was developed. It is the results of this survey which are reported in this paper.

The survey work reported in this paper falls under the general topic of assessment, an area of reinvigorated interest in recent years. As described by Hutchings [**3**], one of the major forces driving assessment has been political pressure from outside the academic community, most noticeably from state legislatures as they express a need for 'accountability'. Overshadowing this source of interest in assessment is the university community's own concern for its students and improvement of its programs. One of the best models of this approach is the Harvard Assessment Seminar [**4**], led by Richard J. Light. The seminar consisted of a number of concerned faculty and administrators at Harvard who shared the results of assessment studies along with teaching and program innovations. The group was soon joined by like-minded people from other institutions, such as the University of Massachusetts, Dartmouth College, and Brandeis University. This model of a campus community of those interested in assessment, experimentation, and innovation in the services of improved teaching and learning is now spreading to a number of different institutions.

This assessment movement deals mostly with outcomes assessment, namely studying the 'value added' to the student or faculty by the college experience. However, as our own experience points out, reform is not likely to be effective without detailed knowledge of the population to be served. Some kind of market study or 'needs assessment' is also called for. Universities have long done these kinds of assessments of their students; what has not been often done is such an assessment at the departmental or course level, relating student characteristics with study habits and course outcomes.

In the mathematical community, a lively reform movement has sprung up and is maturing, as is evidenced by the well-documented interest in calculus reform and the establishment of newsletters such as *UME Trends*. A number of promising approaches have been developed, and many schools are eager to adapt and implement new programs. Our experience at CSUF has pointed out that such reform efforts, to be effective, must be based on a thorough understanding of the current situation. Yet few such studies of student populations have been done or reported, particularly at the large comprehensive schools such as those in the California State system. This survey is the beginning of an attempt to fill this gap.

THE SURVEY

For some time, the author has been interested in the study patterns of her students, doing informal surveys in her classes inquiring into how much they studied, how much time they spent on homework, etc. When the need for a more comprehensive survey was realized, her informal work was developed into the questionnaire shown in Figure 1 (in Appendix 1). The questionnaire was administered during the last two weeks of the spring 1990 semester to students enrolled in the first two semesters of the engineering calculus course and the one-semester preparatory precalculus class. Of 579 students enrolled in the three courses, 150 had withdrawn and were not available for the survey. Of the 429 remaining, 329 participated in the survey. Those students who were too discouraged to come to class or who for other reasons failed to attend for several days were not included in the survey.

Some results were quite surprising while others confirmed notions long supported by anecdotal evidence. The results we are discussing appear in Table 1. As was expected, the percentage of women dropped as the courses became more advanced, while the proportion of engineering students increased. The number of declared mathematics majors was also very low, and does not account for the much larger number of upper-division mathematics students at CSUF. Clearly, most of these mathematics majors are transfer students, rather than being 'home grown' at CSUF. We do not know if the observed number of mathematics majors throughout the calculus sequence is due to students switching into the math major or if it is due to some other cause. The later survey may yield some information.

TABLE 1. Survey results-comparison by class

Item	Precalculus	1st Sem. Calculus	2nd Sem. Calculus
Number of Students	120	109	100
Major (Percent)			
Engineering/Computer Science	34.5	49.0	59.6
Mathematics	0.9	2.9	6.7
Sex (Percent)			
Female	47.0	32.0	36.0
Male	53.0	68.0	64.0
Age			
Median	19.0	19.0	20.0
Mean	20.4	20.4	21.3
Load=Units $\times$ 4 + Worktime+Weekly Commute =Committed hours/week			
Median	75.6	73.0	73.4
Mean	74.2	72.6	73.3
25 percentile	64.2	64.3	65.3
75 percentile	83.3	80.5	82.7
Hours studied per week			
Median	6.0	4.0	6.07
Mean	6.48	4.38	5.82
Six hours or less	64.8%	81.8%	70.7%
Expected Grade			
A or B	37.8%	71.7%	68.0%
C	46.7%	24.6%	25.8%
D or F	15.1%	3.8%	6.2%
Actual Grade			
A or B	37.5%	47.5%	48.9%
C	33.1%	33.7%	28.4%
D or F	29.5%	18.8%	22.7%
Difference between expected and actual grade			
Underestimated grade	12.6%	4.0%	7.0%
Correct estimation	55.9%	43.0%	44.2%
Overestimated grade	31.5%	53.0%	48.8%

Our study did not bear out one of the strongest of the recent demographic trends, namely the rise in the number of older students. Even though there has been a great increase in the number of older 'nontraditional' students at the university, they were not much in evidence in these classes. The median age ranged from 19 in the precalculus and first-semester calculus clssses to 20 in the second-semester course. Students over age 30 accounted for no more than 4% of any of the three classes. 85% to 95% of the students are 25 years old or younger.

The students surveyed do not spend much time studying; they report an average of four to six hours per week study time for their mathematics class. When we consider that these study times are self-reported and so probably somewhat inflated, we are left with a dismal picture. For a four-unit mathematics course, study time of at least eight hours per week is recommended; some suggest study times as high as 12 to 15 hours per week, so the high failure rates of these students should come as no surprise. But in fact, low grades do seem to come as a surprise to many students; on comparing what grades the students predicted to the one actually received, we found that in the calculus courses, fully half the students surveyed overestimated their final grades. In the first-semester calculus course, over 70% of the students expected an A or B, but less than 50% received such a high mark! This is a particularly striking result in light of the fact that the survey was administered during the twelfth week of a 15-week semester, when students already had the results from at least three major examinations and innumerable homework and quiz grades in hand.

These unrealistic expectations also became apparent on examination of the overall amount of time the students were committing to study, work, and commuting. Assuming that for academic success one should study three hours each week outside of class for each hour spent in the classroom (a figure agreed upon by many students of technical majors), we multiplied the student's number of academic units by four and then added on the number of weekly work hours and commute time. This computation gave us an estimate of the number of hours per week the student would need to spend to handle all academic and work commitments adequately. The average student was committed to 73 hours per week, more than 10 hours each and every day, while more than 25% were carrying weekly commitments worth 80 hours! We tend to think that in the traditional college setting, students do not work and live on campus, so commute time is effectively zero. A full-time student carries 15–16 units, yielding a load factor of 64 committed hours per week. This load level is at the 25th percentile among our first-year students!

This pilot survey in the spring of 1990 did not include those students who did not complete the course, namely those who dropped early, withdrew, or had become discouraged. These students are clearly of great interest to those concerned about teaching effectiveness. Therefore, in the fall 1990 and spring 1991 semesters, an expanded survey was given in two parts; during the third week of class when enrollment had stabilized and during the last two weeks of classes. The data from these runs of the survey are being looked at now, and will be described in a later report.

CONCLUSIONS

The results of the preliminary survey already suggest that much useful information with great bearing on reform efforts can and should be gathered. As these data are gathered, the individual departments can make far more informed decisions on the programs that should be offered and how those programs can be designed to be more attractive and successful. Based in part on information from the survey, CSUF has elected to offer the PDP workshop program as a one-unit course, supporting the overloaded students in dedicating the blocks of time necessary for the program. Because so many students work and commute, the workshops are held at midday and in the early afternoons (other schools have been successful with a late-afternoon program).

The age distribution of mathematics students raises a number of questions. Why do older students not enroll in mathematics courses? Should mathematics departments be more aggressive in recruiting from among the group of older students, especially since this segment of the general student population is expected to grow? Given that the demographic pattern of mathematics students is different from that of the general student population, how should student services for them be modified?

We expect that the information in the expanded survey may suggest other actions as we get a clearer picture of the unsuccessful students. Hence, many in the department await the expanded results eagerly, for they have a sense of how useful such information will be to reform efforts. For example, at Fullerton, one useful area for reform might be in the area of freshman orientation and advising. Students may need to be encouraged to carry a light academic load and to work minimum hours for at least their freshman year. More realistic descriptions of the demands of college work may also need to be disseminated at the high school level. Faculty may need to take on a greater role of informal advisers to students, repeating this message yet again.

Many of us at Fullerton were surprised to discover how little information we had about our students, and yet how necessary such information is to improving our department. The results of our preliminary work have already proved useful, and we look forward to finding that the investment of time and effort in the expanded survey will pay excellent dividends. We suggest that such an investment would be useful in any department as it looks ahead and plans for the future.

Addendum

Update to "Student work and study habits at a comprehensive university: A preliminary report". In the time since these preliminary results were reported, the more complete survey was completed and much of the data analyzed. Many of the results already reported were confirmed by the full survey. In precalculus and first-semester calculus, 90% of students were 25 years old or younger; in second-semester calculus, the percentage of those 25 or younger had dropped only to 83–87%. Clearly, the 'nontraditional' student is little in evidence in these classes.

Work and study loads were as high as found in the earlier pilot study. The percentage of students working half-time or more hovered around 50%, from a low of 42% in one class to a high of 55%. A total number of hours of committed time was computed, as during the pilot study, but with a slight modification. Instead of assuming three hours of study per hour of class, we computed 2.5 hours per class hour, yielding a more conservative figure. Still, the numbers were very high, with an average combined work, commute time, class, and study time of 72 hours per week! Reported study times were also low; among students who responded to the second part of the survey (and who were therefore still enrolled in the course), approximately 50% reported studying math five hours per week or less. Students' expectations of course grade still proved to be unrealistic, with the proportion of students overestimating their grades running as high as 50% in some classes.

When we attempted to correlate these work and study loads to the students' grades, we got a surprising result; there was no correlation at all! Conventional wisdom predicts that a heavy work and study load should lead to poorer academic results, but this did not seem to be the case. This result is consistent with that reported by Richard Light's Harvard Assessment Seminars, for his group also found little relationship between workload and overall grade point average. The number of (self-reported) study hours also correlated in some cases only weakly with grade results, and in most classes, no correlation could be found.

The most interesting outcome of this survey was its unexpected usefulness to the mathematics department. For example, a strong correlation was found between scores on a preliminary precalculus skills exam and performance in first-semester calculus. This will be very helpful, as policy on placement examinations is being examined. Data on student persistence and progress through course sequences can also be gleaned from these data. In short, for the first time, the department has a data set to which it can turn for solid empirical information as it engages in discussions of lower-division course policy.

A fuller description of the results reported above and discussion of future research directions will appear in a later report.

APPENDIX

Math 125 1/1 Math Classes Student Survey Spring 1990 CONFIDENTIAL

1. Student ID#__________(for computer analysis purposes, only) 3-9

2. How many units are you taking this semester?__________10-11

3 If you are currently employed, how many hours, on average, do you work in a week?__________12-13

4. How long, on average, does it take you to get to the campus from your usual point of origin? (in minutes)__________14-15

5. In this math class, how many times per week is homework due?__________16

6. On average, how many days prior to the due date do you start to work on your homework assignment?__________17

7. How many times per week do you study for this class?__________18-19

8. On average, how long do you study each time? (in minutes)__________20-21

9. On average, how many hours per week do you study for this class?__________22-23

10 For this class, how many times have you used your instructor's office hours?__________24-25

11. For this class, how many times have you used the Math Tutoring Center(McCarthy Hall, room 187)?__________26-27

12. Are you aware of the workshops available to support this class? Yes_____28 No_____SKIP to Question #15

13. If YES to Question #12: How did you hear about the workshops? (check as many as apply)

Announcement in class__________29
Invitation of instructor__________30
The "grapevine"/other students__________31
Other, specify on back__________32

14. If YES to Question #12: Check the answer below that applies to you, and then check as many reasons below it as apply.

I participate in workshops_____33
Because:
a) It sounded like fun_____34
b) I wanted to work with others_____35
c) I need all the help I can get_____36
d) I like working extra problems_____37
e) Other, specify on back_____38

I participated but then stopped_____39
Because:
a) Too much other studying_____40
b) Work/class schedule changed____41
c) Didn't seem to help_____42
d) Didn't need more help_____43
e) Other, specify on back_____44

I did not participate_____45
Because:
a) No time_____46
b) Work/class schedule conflicts_____47
c) Didn't think I needed it_____48
d) Prefer to study alone_____49
e) Other, specify on back_____50

15. How many times per week do you study with other students for this class?__________51

16. What grade are you expecting for this class?__________52

17. Do you plan to take another math class after this class? Yes_____53 No_____

18. If YES to Question #17: Which math class do you plan to take?__________________________54-56

19. If NO to Question #17: Why not?____

That is the end of the survey questions. Please feel free to make additional comments on the back, and thank you for your response.

© 1991 CSU Trustees

Fig.1

REFERENCES

1. P. A. Treisman, *Study of the mathematics performance of Black students at the University of California, Berkeley*, Mathematicians and Education Reform, Proceedings of the July 5-7, 1988 Workshop CBMath1, Amer. Math. Soc., Providence, RI, 1990, pp. 33–46.

2. R. Asera, *The math workshop: A description*, Mathematicians and Education Reform, Proceedings of the July 5-7, 1988 Workshop, CBMath1. Amer. Math. Soc. Providence, RI,, 1990, pp. 47–62.

3. P. Hutchings, *Watching assessment: Questions, stories, prospects*, Change Magazine, Sept./Oct., (1990), pp. 13–38.

4. R. J. Light, *The Harvard assessment seminars: Explorations with students and faculty about teaching, learning and student life* (First Report), Harvard Graduate School of Education, Cambridge, MA., 1990.

DEPARTMENT OF MATHEMATICS, CALIFORNIA STATE UNIVERSITY, FULLERTON, CA 92634

CBMS Issues in Mathematics Education
Volume 3, 1993

Food for Thought: Applications to Statistics

MARSHA DAVIS

INTRODUCTION

More and more educators of statistics are advocating the use of real-data in the classroom: real-data that students are interested in, familiar with, or have first-hand experience with, provide support for the development of abstract ideas. One of my most successful means of grounding statistical concepts through the familiar and concrete is with food. My favorite food for this purpose is the chocolate-chip cookie.

The nature and sources of variability is the very heart of statistics, giving the discipline its raison d'etre. Just mention the word 'variability' and a myriad of images and connections appears in my mind. But to my students, it is just another word that statisticians use more frequently than the general population—that is, until the bags of cookies arrive. Experiments on chocolate-chip cookies have proved to be a worthy addition to introductory statistics, experimental design, and teacher enhancement courses. Cookies have never failed at in-service workshops for elementary or secondary teachers, programs for high school students, or for my nine-year-old daughter and her friends.

Real-data sets (e.g., numbers of chips in chocolate-chip cookies) provide a context for the numbers involved in an application of a statistical technique. The context aids in drawing meaningful interpretations from the results of an analysis and reduces the likelihood that the computational aspect of the statistical procedure will overshadow the usefulness of the procedure. Mastery of statistical techniques in controlled settings misrepresents the nature of statistics. Real-data seldom conform perfectly to theoretical assumptions (again, variability!) but are inherently 'messy', a condition which often leads to different interpretations. Furthermore, the natural peculiarities of real-data sets require that underlying assumptions of the analysis are reasonably satisfied to insure the validity of the conclusions. If we want to produce responsible users of statistics, we must communicate to students the necessity

©1993 American Mathematical Society
1047-398X/93 $1.00 + $.25 per page

of checking the underlying assumptions on which conclusions are based. This is important even at the most elementary levels.

There are many sources of real-data. Particularly in introductory statistics courses, I have the students collect at least one data set where the experimental units have to be physically handled. Counting chips in chocolate-chip cookies meets this criteria. Data that students collect themselves, by handling, gives the results a tangibility that is not present from data gathered by others or even from student surveys. This physically gathered data has proven to be valuable for grounding abstract statistical concepts covered later in the semester.

The following is a collection of statistical applications involving chocolate-chip cookie data, a collection that has grown steadily over the years. Issues relevant to statistics classes in general precede course-specific suggestions.

When the Chips are Down

Vehicle for introducing terminology. Traditionally, I start many of my statistics classes by walking into the room, arms laden with bags of chocolate-chip cookies of various types and brands. I ask, "What would you like to know about these cookies?" or "What could you find out about these cookies?" (This entrance is a bit theatrical but it immediately alerts students that active participation is required.) Student responses vary from "I want to know how they taste", to "I think Chips Ahoy has the most chips in each cookie. Can we find out if that's true?" (Chips Ahoy always gets picked!) The students' questions provide a good lead-in to the types of questions that could be answered from the material covered during the semester and the types of questions statisticians would, in general, be able to answer.

Below is a brief scenerio of the discussions that typically take place on the first day of class.

Student 1: I think that the Chips Ahoy cookies have more chips than the other brands, especially that store name brand, and I'd like to know if I'm right.

Professor: I'll give you a cookie from each of the brands and you can tell your classmates the answer.

Student 2: Wait! A different cookie from the bag might have a different number of chips.

Professor: Now you're talking about variability. You're telling me that you would distrust your classmate's answer based on a single cookie from each bag because you feel that different cookies from the same brand may have varying numbers of chips. Here we have an example of variability in the cookie-making process.

Student 3: Maybe we could count the numbers of chips in each cookie for the entire bag and then compare the average number of chips per cookie for the different brands.

Professor: I do this experiment every semester. Do you think that your averages will be the same as last semester's class?

At this point the discussion leads naturally into the difference between a population mean and a sample mean. We look at results from previous years and note that the sample averages vary from year to year and from class to class. (Variability due, at least in part, to sampling.) The demarcation between descriptive and inferential statistics is discussed. The initial question about the cookies is rephrased in terms appropriate for inferential statistics.(Students usually decide that they are more interested in brand comparison than in the contents of these particular bags of cookies.)

Thus far, the image of what lies inside the bags of cookies has served as a vehicle for introducing statistical vocabulary: population, sample, population mean, sample mean, descriptive statistics, inferential statistics, variability (due to process, sampling), and bias. (What if one of these bags was from a batch at the bottom of the container and only a few chips got down to the bottom? What effect would this have on our conclusions?) After I began introducing statistical terms in the context of the cookies, I noticed a decrease in confusion over the meaning of the terms. For example, the typical error of writing hypotheses in terms of the sample mean, instead of the population mean, was dramatically reduced.

Reliability of the measurement procedure. The reliability of the measurement procedure is an area that is touched upon only slightly, if at all, in most statistics texts but in practice is extremely important. Pass a single cookie around a room and let each student count the number of chips in that cookie. It is not uncommon for the counts to range from 12 to 28 on a single cookie! With this amount of variability in the measurement procedure, how much validity would you give to conclusions? Generally, a rather heated debate ensues (if the class contains teachers, be ready to turn on air conditioners!) over a standard procedure for counting the number of chips in the cookies. At times, students will demand that the cookies be pulverized so that they can count the exact number of chips in each cookie. (There might, after all, be a glob of chips in the middle that could not be seen from the surface.) A new issue emerges: Must all the experimental units be destroyed in order to obtain meaningful data? Not willing to sacrifice an edible snack, students usually settle on sacrificing two or three cookies to obtain *exact* chip counts for these cookies, only to find that they have to decide what to do with partial chips. No matter how carefully they count the chips, the number they assign is an estimate (with variability). The situation of basing an analysis on estimates is disconcerting for many students. But finally, the class settles on a process for counting. The following is a representative standard procedure for counting chips: count all chips that appear on either the top or bottom of the cookie; chips that appear on the side get counted only once; count chip flecks only if they appear bigger than 1/2 a chip; add two to account for the hidden chips in the middle of the cookie. (Did we really need to add the last step? How would a constant undercount of, say, two chips affect comparisons among brands?)

A second cookie is passed around the room. Students count the number of chips using the newly adopted standard counting procedure. This time there is considerably more agreement among the counts. While there is still variability due to the counting procedure, it has been greatly reduced. However, to keep tabs on the reliability of the estimates for the chip counts, two (or more) independent counts are taken on each cookie. This exercise introduces students in a very concrete way to the necessity of checking the reliability of the measurement device prior to collecting the full set of data. (This is an excellent lesson for future science students: physics students learn the value of calibrating instruments; chemistry students learn the value of using the same scale throughout an experiment; biology students learn the value of standard methods for counting microscopic organisms.)

Introduction of a statistical computing package. When possible, students enter the data into a computer (most of my classes and workshops have access to computer facilities). Minitab is an excellent package for students, requiring minimal start-up time. Using a statistical computing package, the data can be easily manipulated for exploration and analysis. Since Minitab is able to construct visual displays quickly, multiple representations of the data can be studied and compared. Numeric information is but a command away. Students spend their time concentrating on interpretation of the graphic and numeric information.

TABLE 1. A PARTIAL PRINTOUT OF A MINITAB WORKSHEET.

ROW	ahoy1	ahoy2	aroo1	aroo2	big1	big2
1	10	10	16	15	20	15
2	19	19	14	12	21	15
3	15	17	19	17	23	16
4	13	16	13	12	11	8

In this class there were three brands of cookies: Chips Ahoy, Chips Aroo, and Big Chips. Two counts were recorded for each cookie by two different individuals. (Note the proximity of the two independent counts on each cookie for Chips Ahoy and Chips Aroo cookies. For the last brand, it appears that one of the counters is either consistently overestimating or underestimating the chip counts. When you use real-data collected by the class, interesting phenomena occur that can add to the students' learning process and keep the professor on his/her toes.)

Students' discovery of patterns, contrasts, and exceptions can lead to the construction of their own hypotheses. With the aid of the computer, students can tackle problems of a more complex nature. Problems no longer need to be so oversimplified and 'clean' that they lose all connection with reality. Data consisting of two counts per cookie for several bags of cookies would have been very tedious to analyze by hand. Comparative dot plots, histograms, and boxplots provide visual representations of how chip counts differed among the brands and between the two counts.

Course Specific Suggestions

After the initial data collecting activity, I have pursued a number of different directions, depending on which class I am teaching.

Introductory statistics/probability and statistics. It is good statistical practice to precede formal inference with data analysis. As a statistician I subscribe to this practice, and I educate my students accordingly by including EDA techniques in the curriculum of all introductory probability and statistics courses, even mathematical statistics. "Data analysis is most effectively carried out on data with which we are intimately familar, for familiarity suggests both expected features to look for and explanations for unexpected features" [**2**, p. 104]. Students have had years of experience with chocolate-chip cookies. However, little information about the chip counts is gained by looking down long columns of numbers. Data needs to be organized and summarized in order to 'tell its story'. The complexity of the cookie data requires basic exploratory data analysis tools—histograms, stem-and-leaf displays, box-and-whisker plots—to unlock its mysteries. Discussion of appropriate numeric summaries for location and dispersion (mean, median, mode, trimmed mean; range, inner-quartile range, variance, standard deviation) are discussed in conjunction with the visual displays of the data.

Sometimes I have students concentrate on the shape of the cookie data: unimodal or multimodal, symmetric or skewed? The shape of data leads into modeling with theoretical distributions and probability. Does the cookie data look normally distributed? Students construct normal probability plots for the chip counts of a single brand and interpret from this plot how the distribution of cookie counts differs from normality. Perhaps another distribution that the class has studied is more representative of the chip-count data.

Remember those students who wanted to know how the cookies tasted? Their question was reformulated into a question about brand preference. A taste test was set up. Issues of randomization and double blind testing needed to be addressed prior to the collection of the taste preference data. If we assumed that the class responded similarly to a random sample of college students, what conclusions could be reached? (1) Was the indication of brand preference statistically significant? (2) What would be the estimates (both point and interval) of the true proportion of students who prefer each brand?

Reliability of the measurement procedure is important for technical students to understand. When I taught probability and statistics geared to engineering students, we discussed why the average of the independent counts on each cookie tends to produce a more reliable measurement than individual counts. In addition, students use Q-Q plots to compare the chip count distributions for two brands of cookies. The issue of quality control or consistency of the product was evaluated. Sample cumulative distribution functions were constructed and compared to various theoretical c.d.f.'s.

Most importantly, students got to work on a problem that lasted longer than a single chapter in the textbook. They revisited the data set throughout the semester, approaching it first on an exploratory data analysis level and later on an inferential level. In some classes students were given the task of working on an open-ended project of simply analyzing the chocolate-chip cookie data, while for other classes a more directive series of questions pertaining to the cookie data was asked. Having to figure out how to deal with the complexity of the data set in answering questions that did not include an outline of the procedure really tested the students' ability to apply what they had learned. And when the data did not quite fit into the theoretical assumptions of a particular analysis, students had to use some common sense in deciding how to proceed.

Experimental design. Getting students to understand that overall variability in the data can be decomposed into variability due to various sources is difficult. Cookies can be used to make this concept more accessible. Students note first-hand the variability due to the measurement procedure, they know that there is variability due to the cookie-making process and due to different brands. If I bring in deluxe cookies with bigger chips, they expect variability due to the deluxe label even when the brand has remained the same.

Over the years, I have used the cookies to illustrate a number of experimental designs. Here are some ideas. Start with three or more different brands of cookies for a one-way design. A number of brands make a selection of types of cookies, such as deluxe or bite-sized. So add cookie type as an additional factor in the design. But, we also have replications of counts on each cookie taken by different individuals, so add in the factor of replicated measurements. Brands can be broken down into national brands or generic products or store brands. Use your imagination. Finally, just when you thought there was already sufficient complexity, there is the issue of balanced designs versus unbalanced designs. If you choose a balanced design, how do you make the decision of which data to discard?

Now we concentrate on the appropriate measurement for the response. Our data is in terms of chip counts per cookie. But is the chip count per cookie the appropriate measure for comparing brands or types within brands? What if the cookies differ in size? What if the sizes of chocolate chips vary? In either of these cases, what would be an appropriate measure of concentration of chocolate in the cookies? This is a very real issue in an experimental design course; the raw measurements may not be the appropriate response on which to base an analysis.

Next, get ready to carry out the analysis. Perform diagnostic tests to check that the underlying assumptions of the analysis are satisfied. Generally, application of a square root transformation is needed before further analysis is carried out. (This can vary from class to class, but is not surprising since we are working with count data.)

The concrete example furnished by the cookie data has supported a number of fundamental concepts that are often difficult for students to grasp. While counting chips in chocolate-chip cookies may seem irrelevant, I remind my students that the process of counting chips has parallels to counting bacteria in a culture dish or red blood cells on a slide (except chips do not move).

Teacher enhancement: "Conceptual foundations of statistics". * For ten years, the SummerMath for Teachers (SFT) program at Mount Holyoke College has introduced teachers to a constructivist approach to mathematics instruction, a philosophy of education that is very much in harmony with the vision presented in the NCTM's *Evaluation and Curriculum Standards.* Included in the *Standards* is the recommendation that probability and statistics be represented as key content stands in the K-12 mathematics curriculum. "Conceptual Foundations of Statistics" was developed as a semester course under the auspices of the SFT program to both model a constructivist teaching methodology and to supply content background in statistics. (For more detailed information on the course or the SummerMath for Teachers Program, see [**1**].)

For this course, most of the lessons were based on data sets collected by the class. Teachers posed questions which arose naturally from the context of the data or from patterns that teachers discovered within the data. These questions were instrumental in setting the direction for the class. Care was taken to select data-gathering situations and required reading material that would introduce a cohesive collection of statistical concepts over the course of the semester.

Again, data from chocolate-chip cookies provided the initial and recurrent backdrop for statistical discussions. At first, teachers decided that they wanted to find out which of the three brands sampled had the highest average number of chips per cookie. They collected two independent counts on each cookie and, in addition, were given access to a similar data set from an undergraduate class. (One difference in the two data sets is that the undergraduates worked with small bags of cookies while the teachers had larger sized bags.) After sifting through the data, via comparative histograms and boxplots, the teachers posed additional questions concerning the chocolate-chip cookies. For the next several weeks we worked on statistical methods that would aid the analysis of the following topics.

---

*The material in this section is based upon work supported by the National Science Foundation, under grant number TPE-8850490. Any opinions, findings, conclusions, or recommendations expressed in this material are those of the author and do no necessarily reflect the National Science Foundation.

(1) Reliability of the counting procedure.

(2) Characterization of the distribution of cookie counts within a single brand of cookies.

(3) Comparison of chip counts for the three brands of cookies.

(4) Comparison of the undergraduates' data versus the teachers' data for reliability.

(5) Comparison of results based on the teachers' data with results based on the undergraduates' data. (Would both data sets lead to the same conclusions?)

(6) Combining information from the two classes (teachers and undergraduates) to get better (?) results for #3. Given the results from #4, was it reasonable to combine the two data sets?

(7) Consistency of each product.

Topics three and seven forced a discussion of measures of location and variability. Teachers already knew how to compute the mean median and mode of a data set. Because teachers could compute the mean but did not have a clear concept about what the mean measured, we spent some time on concrete representations for the mean. The issue of selecting an appropriate measure of location for the cookie data was eye opening for many teachers. Measures of variability were even more elusive. The teachers kept asking for information on the standard deviation. The computer could and did provide it and all the teachers had computed standard deviations in a previous course, and yet they still had no clue as to what it meant. Five data sets involving class exam grades (familiarity!) were supplied and the teachers were assigned the task of creating appropriate measures of grade variability for the classes. The teachers invented the range, essentially the inner-quartile range and the mean absolute value of the deviations of the data from their mean (MAD). The latter measure provided the linkage needed for the teachers to understand the standard deviation. After these discussions, the teachers were better able to comment on topics three and seven.

During the time teachers were working on these questions, I assigned several articles that contained discussion on the concept of a weighted mean [e.g., **4**]. During the data analysis, teachers were stumped on how to calculate the overall mean from the two classes' sample means. (Since the teachers and undergraduates had worked on samples of substantially different sizes, due to the different sized bags of cookies, it was inappropriate to simply average the two classes' averages.) After some discussion the teachers made a tie-in to the articles and were able to construct the overall mean by weighting the sample means of the two classes.

The impact of this course, including the usefulness of the cookie data, is illustrated by an experience related by one of the participating teachers. She was the subject of a research project which evaluated teachers' interaction with their students. Data was collected as to the number of times teachers interacted with their students and the nature of that interaction. After viewing the final report, this teacher commented that "Conceptual Foundations of Statistics" gave her the background and confidence to ask the researcher questions about the reports' conclusions. Because teaching styles had not been taken into account, teachers had not been observed for the same lengths of time, and student absenteeism had not been incorporated in the interaction measure, she questioned the validity of the report. The teacher wrote, "I kept thinking about the cookie data with all its sources of variability....I couldn't believe the numbers [supporting the researcher's conclusion] meant anything."

Conclusion

In all of the situations described above, something as simple and familiar as chocolate-chip cookies have supported a deeper understanding of a diverse collection of abstract statistical concepts. I hope my experiences and suggestions have provided 'food for thought'.

References

1. Marsha Davis, *Conceptual foundations of statistics: A constructivist and technological approach*, Doris Casey, et. al., ed. Technology and Teacher Education Annual, School of Mathematics, East Carolina University, Greenville, North Carolina, 1991, 355–358.

2. David S. Moore, *Uncertainty*, in Lynn A. Steen (Ed.) On the Shoulders of Giants, National Academy Press, Washington, D.C. 1990, 95–138.

3. National Council of Teachers of Mathematics, *Curriculum and evaluation standards for school mathematics*, Reston, Virginia, 1989.

4. A. Pottatsek, S. Lima, and A. D. Well, *Concept or computation: Students' understanding of the mean*, Educational Studies in Mathematics, **12**, 1981 191–204.

Eastern Connecticut State University, Willimantic, Connecticut
E-mail address: davisma@ecsu.ctstateu.edu

CBMS Issues in Mathematics Education
Volume 3, 1993

Report on *Geometry and the Imagination*

JANE GILMAN

INTRODUCTION

During the last year and a half, I have been involved in developing and teaching two innovative mathematics courses. Both courses were designed to address a variety of issues of current concern in mathematics education, although each course was geared toward a different population. The main thrust of each course was to develop both a new syllabus and new teaching methods.

Both courses placed a strong emphasis on exposing students to the process of thinking about mathematics. Formal lectures were avoided as much as possible. Students were encouraged to *discover* mathematical ideas and concepts by breaking the class into small discussion groups to consider problems. The motivation for this was two fold. First, there was a belief that a concept which a student has discovered him or herself is one which is truly understood so that it can never be forgotten or unlearned. Second, there was a belief that the beauty of mathematics and the pleasure of doing mathematics, which is what we should be sharing with students, is intimately connected with this process of discovery.

The syllabus developed needed topics that lent themselves readily to such a type of discussion format and to such a teaching approach. While the topics had to be inherently of mathematical interest, they also had to be accessible: that is, topics where one could get to conceptual and theoretical questions without developing a long technical background.

THE TWO COURSES

Both of the courses were created and taught by teams. One course was entitled *Geometry and the Imagination* was first taught at Princeton University

This paper was originally a talk given at the AMS Special Session on Mathematics and Education Reform, Baltimore, 1992.

The author is partially supported by NSF grant #DMS-9001881.

©1993 American Mathematical Society
1047-398X/93 $1.00 + $.25 per page

by John Conway, Peter Doyle, and William Thurston. A second version was also given at Princeton by Conway, Doyle, and myself. An intensive two-week version was given this past summer by all four of us at the University of Minnesota Geometry Center.

The second course, entitled *Fresh Honors Mathematics*, was given at Princeton primarily by Bill Thurston and me, but with significant input from Peter Doyle.

Fresh Honors Mathematics was for potential mathematics majors and used dynamical systems and iteration as a setting for reviewing some of the more theoretical material from Calculus I and for integrating the use of computers into the curriculum.

Geometry and the Imagination served a broader population, both majors and nonmajors. However, the nonmajors were expected to be students who thought of themselves as 'liking' mathematics.

In this talk, I will mostly concentrate on describing the experience this summer when *Geometry and the Imagination* was taught by the four of us in an intensive two-week format. We spent three hours every morning in the classroom. Students spent the afternoons working on projects in what was called the 'geometry room', a room where there were many geometric models and which was always staffed by at least one mathematician willing to talk about any and all questions. There was a diverse group of students. The fifty students included gifted high school students from the neighboring area, high school teachers and college teachers from the neighborhood, and undergraduate mathematics majors from colleges across the country.

A Typical Day

A typical class day would begin with the four of us entering the classroom with one of us dragging a wagon full of objects which we were likely to use that day: polyhedran, mirrors, flashlights, vegetables, what have you. The class would begin with a brief lecture, as brief as possible, outlining a few basic facts and posing a problem or a set of questions for discussion. The class would then break up into discussion groups of three to five people and consider the questions. After an appropriate interval, the class would be reconvened and one of the four of us would do what we called 'taking the discussion'. Each group would have a reporter, the reporters rotated, and each reporter would share his or her group's findings with the class. This would sometimes lead to formal conclusions and other times lead to further discussions which took various forms.

Sometimes they were discussions among the class as a whole. Other times they were discussions among the four of us which took place in front of the classroom. Yet other times they were discussions that required the class to break up into small groups again to consider further questions.

Team Teaching

While the course was team taught, it was not done in the conventional mode. When I mention team teaching to colleagues, I am often asked who taught what. The answer is that we all taught all parts. Usually we were all in the classroom at the same time, interrupting each other, contradicting each other, and sometimes helping each other. Our interactions were often lively.

A number of times after I finished explaining some point or concept to the class, I would back away from the front of the class and notice that one or even two of my colleagues had been drawing pictures and diagrams on the blackboard to illustrate my point or they had improvised an illustration behind me using some of the props from the wagon. It felt strange this fall to return to a regular classroom. I would finish explaining a point or concept, walk away from the front of the room, and turn, expecting to find a nice diagram on the blackboard, but there was none.

The blackboard in the room in which we taught in Princeton had three panels: the ones that slide up and down. At one point I looked around the classroom to see that each of the three of us was simultaneously standing, chalk in hand, in front of a portion of the blackboard trying to explain *our* perspective on the issue at hand to the class of students. It was shortly after this episode that a student came to the geometry room one afternoon and told us that he had begun telling his friends that he was not being team taught, he was being gang taught.

Hands-on Materials

A major ingredient in the course was using the props or hands-on materials. Here are some examples.

There was a great deal of discussion of symmetry and reflections leading eventually to a surprisingly sophisticated discussion of quotient orbifolds and fractional Euler characteristic. This was initiated by passing out fifty pairs of 4×4 mirrors and asking:

How do you hold two mirrors so as to see an integral number of images of yourself?

Now, one can stand up in front of a class and explain that one can hold two mirrors at an angle and that as one changes the angle one will see an integral number of images for some angles but not for others, and the students will believe it, but it has a much bigger impact if the students discover it for themselves. In the process of discovering it and playing with the mirrors, they will also notice other phenomena, some of which will be articulated in the ensuing discussion, others of which will not be, but which will remain as passive knowledge that will facilitate further discussion.

This is a teaching technique about which I was initially skeptical. It seemed extravagant, even silly, to use so many mirrors. However, I became a believer as I saw the process work and saw the concrete examples lead to interesting and substantive theoretical discussions.

Here is another example: the vegetable example. Take a potato. Peel a strip, a closed strip about the equator. Lay the strip flat out on the table. Measure the angle by which the peel falls short of closing up to be a full circle or exceeds a full circle. This measures Gaussian curvature — the Gaussian curvature of the region of the potato enclosed by the peel. Now take other closed peels, not necessarily equatorial peels, and other vegetables, cabbage and kale, for example. Again, this simple exercise became the starting point for a surprisingly sophisticated discussion and understanding of curvature.

I also learned why artificial flowers look fake. Flowers and leaves have lots of curvature. Look at how curly or negatively curved kale is. Artificial flowers are usually made out of flat pieces of paper or material. The lack of curvature is why they look wrong.

Incidentally, this exercise was less extravagant than the mirror one, for we felt obliged to eat the unused cabbage and kale for dinner that night.

Team Teaching: A Creative Endeavor

The team teaching required extensive interaction outside the classroom. We would gather after class to review the previous class and to plan the next class. We would discuss what material had worked well and why, what had not worked well, which concepts had been easy to get across, which had not, how each of the diverse groups in the audience were faring, and other similar issues. We would also begin to plan the next class, discussing such issues as what topics should be covered, what discussion questions should be posed, what notes needed to be written, and what props would be needed.

These discussions were quite lively and frequently heated as the four of us held very strong views about mathematics. Sometimes it seemed that the discussions never ended. If we quickly reached agreement about the broad outline and objectives of the next class and ended the planning session early, it was dangerous to run into each other again that day for we might discover that someone had had a change of mind. Sometimes we were still arguing as we walked into class.

Our disagreements were often as productive as our agreements. We could not reach an agreement about which of two proofs of Descartes' angle defect formula to give. One proof suggested was a straightforward and direct computation, a proof which some of us thought the students might feel that they themselves could have come up with. The other proof was more conceptual and prettier, but unmotivated. Since we could not agree on which to present, we presented both of them and focused the discussion questions on what were the relative advantages and disadvantages of the two proofs. In fact, in the course of the class we gave a third and possibly even a fourth proof.

It was very interesting a few days later when we read the students' journal comments about that class. I expected to learn which proof had won the popularity contest. But instead students wrote that they had never known that there could be more than one way to prove something. They thought that math was cut-and-dried and they had never imagined that one could discuss which was a better way to prove something or which was a prettier way.

CONCLUSION

In summary, I would like to say that the course was a lot of fun and a lot of work. The students seemed to respond well and to learn a great deal. The methods led to some deep discussions.

A number of natural questions arise about the course. For example, can one adapt these techniques to a situation where a certain fixed amount of material has to be covered? What happens if the course is taught with fewer resources, either by fewer persons or with a larger class or without mirrors? During this academic year, some of the people associated with the course are teaching either *Geometry and the Imagination* or other courses covering different topics but using similar techniques. The courses are being taught under a variety of different circumstances at a variety of institutions. The experiences of this year may give us answers to some of the questions.

MATHEMATICS DEPARTMENT, RUTGERS UNIVERSITY, NEWARK, NJ 07102
Current address: DEPARTMENT OF MATHEMATICS, RUTGERS UNIVERSITY, NEW BRUNSWICK, NEW JERSEY 08903
E-mail address: gilman@andromeda.rutgers.edu

CBMS Issues in Mathematics Education
Volume 3, 1993

Language Acquisition and Mathematics Learning

NANCY CASEY

Nature of Language Learning

Language is a beautiful, complex, all-encompassing human activity that requires no special gifts. We all do it. Language is not taught to people, it develops within them. Barring brain damage, human children learn whatever language or languages to which they are exposed.

The richness of a child's linguistic comprehension and expression is directly related to the richness of the linguistic environment. Children learn the languages that they hear. If, in addition to conversation, they are read a wide variety of literary forms—stories, poetry, nonfiction—they learn the language of those forms as well. When children are exposed to the ways that others express themselves in writing, they will recognize those forms, make their own 'sense' of how they are used to communicate, and attempt to express themselves in similar forms. The child's expression and comprehension is creative, not strictly imitative. Children do not simply reproduce the forms and vocabulary to which they are exposed, rather they adapt them for their own communicative needs.

Extending Language: Learning to Read and Write

In the mid 1970s, research in cognitive and developmental psychology and linguistics was brought to bear on the issue of children's learning to read and write. Reading was recognized as a psycholinguistic process in which a person's knowledge of the world, understanding of language, and expectations for a text are all engaged in the process of deriving meaning from the printed page. Knowledge of phonics, sentence structure, and rhetoric compose only a small part of the psycholinguistic awareness which makes the written word meaningful to the reader. A child's development as a reader is the development of a larger ability to make predictions about the meaning of the writing and either confirm or refine them with information they obtain from the page.

©1993 American Mathematical Society
1047-398X/93 $1.00 + $.25 per page

It became clear that written language is a communicative form unto itself and not simply an encoding of the spoken word. Approaches to teaching reading and writing that relied solely on mastering rules of sound-letter correspondence were too narrow in focus. If the learning process for written language is the same as that for spoken language, then it is governed by the child's inherent motivation to make sense of the world and to communicate, and it is best enhanced by rich exposure in meaningful context. For nearly 20 years, teachers and researchers have been seeking to understand how reading and writing operate within the language-learning processes that are already very active for young children.

The development of individual children's ability to express themselves, communicate, and understand others via writing is, on the one hand, unique for every child, and on the other, tends to follow a general progression for all children. At the core of understanding and using written language is the notion that marks on paper make sense or carry meaning. As the child develops cognitively, so does the depth and complexity of the meanings he can associate with the marks. In addition, the child gradually internalizes concepts about how writing is organized: reading from left to right, page ordering, chapters, poetry versus prose, the way ideas progress in a story, etc.

Just as speaking and understanding are interconnected and learned together, so too are reading and writing. A child who knows that meaning can be taken off of a page will attempt to put meaning on a page with writing. Writing emerges from drawing, as a child draws letters as pictures, engages in 'pretend writing', and puts ideas on a page with letters and pictures. Gradually, sound-letter correspondences and other spelling conventions internalized from reading are actively used in writing with increasing accuracy. When freed to do so, children's motivation and desire to write are fueled by their drive to communicate.

Of greatest importance, all children, when given an opportunity to express themselves creatively in such an environment, reveal themselves to be deep and creative thinkers. Their meaning-making activities with words on a page are but a subset of their continuous endeavors to make meaning of everything they see and know. They are curious, fresh, and ruthless observers of their environment. They seek patterns, make generalizations, and just as eagerly delve for exceptions and counterexamples. Opening the door to writing has provided children with many opportunities to analyze their world, to ask questions, and to communicate with others about what they are thinking. The door to mathematics can be opened for them in the same way with equally exciting possibilities.

Mathematicians describe and recognize patterns, make abstractions about those patterns, draw conclusions about the natural world, or manipulate pure abstractions out of sheer joy. Drawing upon a common vocabulary, notation, traditions, and body of knowledge, mathematicians communicate with one another about their perceptions and insights. The language they use is mathematics. Children who are invited to use that language to communicate in meaningful ways about their perceptions of the world will blossom as young mathematicians, just as they blossom as readers and writers when the full communicative power of written language is made available to them.

Many of the marks that children make on paper are marks that they have seen. What if they are exposed to mathematical symbols to the same extent that they are exposed to letters? What if someone talked to them in the language that those symbols represent, read to them from texts that contain those symbols, and used the language (the words, the concepts, the symbols) to talk about the concrete world around them and to share abstractions? Would not children discover in mathematics a language for exploration, play, and communication that opens otherwise inaccessible doors to insight and discovery? Might not this language also provide them with facilities to describe ideas they already have for which the language of daily chatter and literature are found wanting? Would we not discover that the line that divides mathematics from language arts is a language barrier? If so, extending children's range of expressive and comprehensive abilities to include the things that mathematicians talk about will dissolve that barrier.

Classroom Practice: Encouraging Language Development in a Language-rich Environment

The matter of teaching reading and writing is that of enhancing language development, and the most effective way to do this is to enrich the environment linguistically. This is not done haphazardly. The teacher's responsibilities in the 'whole language' curriculum are to command a deep knowledge of the subject matter, understand the child's learning process, provide a language-rich environment, and invite children to participate in a larger community of literate people. The result is a student-centered orientation to teaching and learning. The teacher's focus is each student's development; the student's motivation for learning comes from within.

The teacher's depth of knowledge provides the flexibility that student-centered learning requires. A language arts teacher is well-read across many genres, forms, time periods, and themes and can frame a concept, not in a single, logical manner, but in whatever manner the situation, a question, or a student's receptivity requires. The teacher is a wealth of suggestions, encouraging students individually and collectively to follow tangents: a similar form, another slant on the subject, a view from a different century, an invitation to creative expression. The direction that students' learning takes is never limited to the ordering imposed by a textbook sequence of exercises and activities.

The structure of the learning is rooted in the teacher's understanding of the learning process—for the group of children as a whole and each child individually. The teacher decides how to proceed, what questions to ask, and what encouragement to offer by communicating with students about what they are learning. The teacher tries to understand where the students are and to bring them forward.

The 'language-rich environment' is the framework within which the teacher can do all of this. The children have much to talk and think about, interesting and stimulating things to read, meaningful and authentic contexts for writing, and the freedom to develop. Available or close at hand are books, posters, pictures, anthologies, fiction, nonfiction, poetry, reference materials, a wide variety of writing and book-making materials, film, videotape, tape recorders, and so on. The classroom walls are a mosaic of the creativity and inquisitiveness of the collective mind of the group: artwork, posters, writing, questions.

The order and organization in such a classroom exists at a level other than that of regimentation. It is the responsibility of the teacher to know what is going on, who is learning and working on what, and how it is going. To arrive at this point, the teacher must teach the children how to accept the responsibility for their own learning, how to behave, how to follow up on their own inquisitiveness, and how to learn from and help one another.

These classrooms are veritable beehives of activity and, depending on the moment, can seem chaotic and disorderly at first: students bustling, talking, discussing things, while the teacher moves among groups and individuals, observing, listening, engaging in thoughtful communication with them. When the teacher does move to the front of the classroom to command the attention of the group as a whole, it is to present something relevant to everyone, and it is not at all unusual for the 'teacher' to be a student.

The Community

As users of language and wielders of pencils, all humans are writers. The most effective writing teachers accept themselves as writers. Levels of skill and passion vary among writers as among human beings, but from anywhere on the continuum, writing teachers who are also writers can share with their students the joy and power of writing. They impart facts and teach skills that are useful to the child at the child's own point of development as a writer. As writers, they extend to a child a reasonable and genuine invitation to join and to be active in a larger community of literate people.

As recognizers of patterns and manipulators of abstract concepts, are we not all mathematicians as well? Language arts teachers were sometimes appalled and daunted to discover that to teach writing they had to become writers. The seemingly insurmountable task became a joyful one when the real requirement turned out to be developing an inherent ability. It is time to begin encouraging mathematics teachers to accept themselves as mathematicians— to learn the language by using its vocabulary to name what they see and understand, to see the world through the lens of the new concepts that fluency in the language gradually opens.

Language arts teachers are becoming more experienced and comfortable with dealing with writer's block. It is never terminal. Find your voice, and it always has something to say. With guidance and encouragement, anyone can learn to free the writer within, a writer that has often been imprisoned by earlier failures to jump through the hoops of so-called writing instruction. Are not similar failures the source of math anxiety, too? If so, the cure is in encouragement, guidance, and success in using mathematics to communicate in a meaningful way with other mathematicians.

What are the elements of the mathematics-rich environment? What do you add to the language-rich environment to include the language of mathematics? The answers to these questions will come from creative experimentation by teacher-mathematicians as they try new things and observe how their students respond to them. There are many possibilities:

(1) Make the symbols accessible and available as models and inspiration for drawing, just as letters are.

(2) Discuss mathematical symbols as a shorthand and as codes for words and phrases.

(3) Make representing ideas symbolically open-ended; encourage children to name concepts and use symbols to represent them.

(4) Model and encourage writing about mathematics. Use journals or folders to store thoughts about mathematics, from theorems to playful whimsy, using words, symbols, and pictures. Create opportunities for children to share their mathematical writing with each other and respond to each other.

(5) Encourage, model, and create opportunities for students to share their work and ideas with mathematicians outside the classroom.

(6) Publish student mathematical writing in school publications for students, parents, the community, and in special mathematics publications that grow out of this process. Make posters, handbills, video and audio tape.

The Teacher/Researcher

It is impossible to propose more than a framework for approaching mathematics learning as language learning. Like their counterparts in language arts teaching have done, mathematics teachers must elevate themselves and claim the titles of mathematician and researcher. (Interestingly, in the elementary classroom, the language arts teacher and the mathematics teacher is usually one and the same person!) As mathematicians and researchers, teachers are guides for and observers of children's development as mathematicians. The mathematics community is relying on them to begin to systematically take risks with their mathematics teaching, plug into students' learning, and report their conclusions to the mathematics community at large.

What will emerge is not a methodology, rather an attitude about and approach to teaching that empowers students to take charge of their own learning. Young mathematicians who leave elementary schools taking this approach will enter high school and college mathematics programs as mathematicians. They will have insights into their own learning processes, and mathematics courses, regardless of 'quality', will be opportunities to enrich their own mathematical power.

For Further Reading

Dave Baker, Cheryl Semple, and Tony Snead, *How big is the moon: Whole maths in action*, Heinemann, Portsmouth, NH, 1990.

Jennie Bickmore-Brand (editor), *Language in mathematics*, Carlton South, Victoria, Australia, Austrailian Reading Association, 1990.

Lucy McCormick Calkins, *The art of teaching writing*, Heinemann, Portsmouth, NH, 1986.

Elliot Eisner, *The enlightened eye: Qualitative inquiry and the enhancement of educational practice*, MacMillan, New York, NY, 1991.

Kenneth Goodman, *What's whole in whole language?* Heinemann, Portsmouth, NH, 1986.

Magdalene Lampert, *When the problem is not the question and the solution is not the answer: Mathematical knowing and teaching*, American Educational Research Journal, 27 (1), (1990) 29–63 (includes extensive bibliography).

Language arts, Journal of the National Council of Teachers of English, Urbana, IL.

John S. Mayher, Nancy Lester, and Gordon M. Pradl, *Learning to write, writing to learn*, Boynton/Cook, Upper Montclair, NJ, 1982.

L. S. Vygotsky, *Thought and language*, MIT Press, Cambridge, MA, 1962.

——, *Mind in society: The development of higher psychological processes*, Harvard Univ. Press, Cambridge, MA, 1978.

College of Education, Washington State University, Pullman, Washington 99164
E-mail address: mfellows@csr.uvic.ca

CBMS Issues in Mathematics Education
Volume 3, 1993

Computer Science and Mathematics in the Elementary Schools

MICHAEL R. FELLOWS

ABSTRACT. Computer *science* is fundamentally about algorithms, recipes for solving problems, and performing tasks. In the same way that children can learn about dinosaurs without digging for bones and about planets and space without peering into telescopes, the intellectual core of computer science is not dependent on machines for its presentation. Just as with these other subjects, an approach based on stories, activities, and ordinary materials can be more vivid and engaging than approaches that make a fetish of computers. We argue that algorithmic topics are a good source of material with which to provide for children in the elementary grades a broad, exciting, and active introduction to mathematics. Our experiences sharing some of these topics with classrooms in grades one through four (ages five to nine) are described. We propose that principles of language acquisition should be applied to the teaching of the mathematical sciences and review how these principles have previously been applied to the teaching of reading and writing. We discuss some of the important aspects of the mathematics research community experience and explore ways in which this experience can be fostered in the classroom. Some kinds of mathematical research in algorithms and combinatorics are actually accessible to elementary-age children, and conversely, interaction with children can sometimes inspire research questions. We describe some examples of this surprising research community.

1. INTRODUCTION

This paper describes and analyses the experiences of the author in presenting a variety of topics in computer science and discrete mathematics to elementary school children of ages five through nine. Some of these presentations were made with the collaboration of Nancy Casey, whose work is described elsewhere in this volume.

The discussion that is offered here can be summarized as follows:

• The competencies required for the increasingly computerized world are essentially mathematical. It is a serious (and common) mistake to make a fetish of the machines.

• Computer science is not about machines in the same way that

©1993 American Mathematical Society
1047-398X/93 $1.00 + $.25 per page

astronomy is not *about telescopes.* There is an essential unity of mathematics and computer science.

• The intellectual core of computer science can be presented to children even in situations where there are no computers (for example, in countries or school systems that cannot afford them), laying a foundation for later computer science education. Many of the core ideas of computer science are best introduced without machines.

• Computer science represents a tremendous flowering of mathematics. It is particularly good news for children because it is a treasury of accessible, colorful, and active mathematics. For introducing children to mathematical science, it is unmatched in these terms by any other source. Think of computer science as the modern 'geometry,' but a thousand times more vivid, varied, engaging, and open to exploration.

• The teaching of the mathematical sciences should follow the lead of, and be integrated with, the 'whole language' paradigm in the teaching of language and writing skills. Mathematics that is rich with stories and opportunities for active exploration is well suited to this language acquisition point of view.

• In the same way that children's art is interesting *as art* and children's writing is interesting as writing, mathematics with children can be interesting as mathematics. There are kinds of research activity accessible to children, and interaction with children can be stimulating for people active in research or at higher levels of learning.

The organization of this paper roughly follows the story as it unfolded, first presenting material in the classrooms and later attempting to gain a broader understanding of how these experiences fit into the larger field of current discussions in mathematics education. The classroom experiences came first, because my involvement began simply as a 'parent volunteer' contributing classroom topics for an hour or an afternoon.

Section 2 describes (retrospectively) the objectives of these classroom presentations. Section 3 provides some details of the topics and activities which were brought to the classrooms. Section 4 concerns the subsequent effort to relate these classroom experiences to current discussions in the field of mathematics education. Section 5 describes a supportive point of view for these activities in language arts education. Section 6 discusses how classrooms can function as research communities. In the last section, some mathematical research problems that were inspired by the classroom presentations are recounted.

2. The Objectives of Our Classroom Projects

There seem to be at least two fundamental problems with education in the mathematical sciences in grades one through four.

(1) Most children in these grades are never exposed to mathematics. Arithmetic is not mathematics!

(2) Most children in these grades are never exposed to computer science, despite all the personal computers in the classrooms. Programming is not computer science!

In contrast, children in these grades are often exposed to the central questions and activities of geology, astronomy, biology, chemistry, etc. They are sometimes exposed to the frontiers of knowledge in these subjects, as exciting recent discoveries and developments are discussed in class. They are exposed to art, music, and literature, and their creative efforts in all of these areas—their writings, art, and science projects—are valued.

There is a tendency to apply miserly, and mistaken, standards of 'real world' concern to the curriculum of the mathematical sciences that are not applied to any other subject. It is not approached as the playful, fascinating, and beautiful enterprise that it is, competitive with dinosaurs and outer space. It is usually treated rather as the necessary dreary accumulation of skills for someday balancing checkbooks and figuring mortgages. What life-skill needs do the subjects of dinosaurs and outer space address? If the exposure of children to literature were similarly limited to tax forms, job applications, and parking regulations, then reading would be as widely loved as mathematics is today.

Mathematics, the language of science, and its principal modern branch, computer science, can be presented to children in these grades in wonderfully engaging and active ways, emphasizing their role as the language of science and technology. Children can be presented from the beginning with the essential unity of mathematics and computer science. Mathematics presented as a research enterprise can also provide fair opportunities for children to be shown that the world is full of questions to which adults do not know the answers.

The central questions of computer science are conceptual, and appreciating this science does not depend on sitting in front of a terminal any more than appreciating the questions of astronomy depends on holding your eye to a telescope. Similarly, the competencies that are most important for coping with an increasingly computerized world are essentially mathematical. Programming and 'experience with computers' are relatively unimportant in contrast to mathematical literacy and confidence in mathematical modeling and problem-solving.

In many school situations where there is not a machine for every student and there are scarce opportunities for using the machines, making a fetish of computers may have the further negative effect of increasing the disadvantage of female and minority students who tend to lose these opportunities to pushier cohorts.

Computer science is thoroughly permeated with discrete mathematics. Together these subjects constitute a fertile source of accessible, colorful, and concrete problems for presenting mathematical modeling, reasoning, and open-ended exploration. For the early school years, these subjects are unmatched in this regard by any other kind of mathematics. Computer science adds to this richness, as it does to mathematics in general, by highlighting and elaborating the issues of computational activity and resource economy. These intertwined subjects constitute a modern treasury of accessible, active, and applicable mathematics.

The goals of the visits to elementary school classrooms can be summarized as follows:

- To show that mathematics is fun and full of stories, activity, invention, and play.
- To show that mathematics, like dinosaurs and outer space, is a live science with visible frontiers of knowledge.
- To present the essential unity of mathematics and computer science and display the intellectual core of the latter.

Working with young children is interesting for several possible reasons. First, lasting attitudes towards mathematical science may be formed in the earliest grades. Second, competition with the deadly traditional 'school mathematics' curriculum is less of an issue in these grades, and the deficiencies of the traditional curriculum are more starkly apparent. Third, if interesting 'college level' topics (such as the Muddy City problem described in the next section) can be made accessible and interesting to second-graders, then these topics will also be accessible at intermediate levels.

The outlook of the author is that of an active researcher in mathematics and computer science, and the story that is recounted in this paper is basically that of an enthusiastic, intellectually naive adventurer in the world of elementary mathematics education.

3. Some Math and Computer Science Topics for Young Children

The purpose of this section is to describe some of the topics which were presented during the classroom visits. These visits were to middle-class classrooms of children aged six to ten in Moscow, Idaho, and Victoria, British Columbia during the period of time September 1989 to June 1992. The typical format for a presentation was a 60-minute block of time in which to present the topic and organize activities and discussion. In one of the classrooms (second grade) the children kept individual mathematics journals.

The Muddy City. The technical name for this topic is the problem of computing a *minimum weight spanning tree* in a graph. Several efficient algorithms for solving this algorithmic problem are known and are routinely covered in any college level course on algorithms and complexity [**CLR**]. The story presentation of the problem next described is meant to be entertaining, but it should be noted that there are many practical applications of this problem in computing. This is true for many combinatorial optimization problems; they support the modeling of both fanciful and industrial situations.

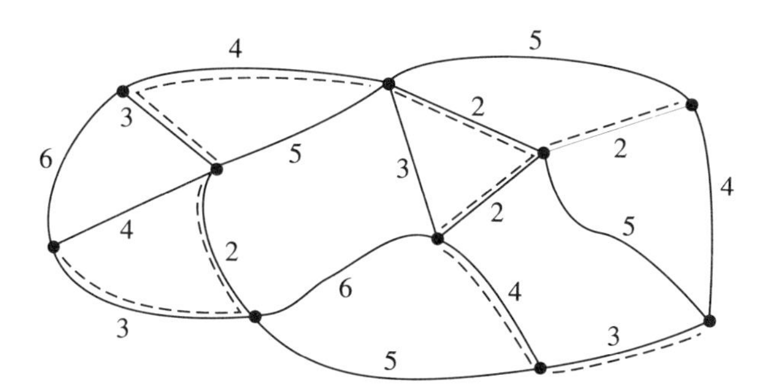

FIGURE 1. The Muddy City and an optimal solution.

The children are given a map of the Muddy City (see Figure 1) and told the story of its woes: cars disappearing into the mud after rainstorms, etc. The mayor insists that some of the streets must be paved and poses the following problem. (1) Enough streets must be paved so that it is possible for everyone to travel from his or her house to anyone else's house by a route consisting only of paved roads, but (2) the paving should be accomplished at a minimum total cost, so that there will be funds remaining to build the town swimming pool. The number associated with each street in the map of the town represents the cost of paving that particular street.

Thus the problem posed is to devise a paving scheme meeting requirement (1), connecting the town by a network of paved roads that involves a minimum total amount of paving. The cost of a paving scheme is calculated by summing the paving costs of the roads chosen for surfacing.

Some of the five-year-olds began by figuring out where the new town swimming pool should be located and which node represented their house! In general, posed with a problem of this sort, the classroom explodes with activity, and there is a tremendous range of response. Some students rapidly understand the problem, while others require further explanation as they consider partial solutions and examples. (It is always useful to delegate to those who understand the problem of job of explaining it to others as tutors.) One fascinating aspect of the classroom experience was the reports of the teachers that their expectations concerning student performance were turned topsy-turvy; the children who did well on these problems were not always the same as the ones who did well at the usual arithmetic drill.

The children worked on the problem, usually in small groups, with the immediate objective of finding the best possible solution. This was typically recorded in a place that everyone could see. Students were asked to describe their strategies and ideas, both as they worked and in a concluding discussion. In classrooms where the students kept mathematics journals, they also wrote descriptions of the problem and of their ideas on how to solve it. These mathematics journals were instituted with great success in a (latter part of the year) second-grade classroom and in a fourth-grade classroom.

As part of the wrap-up discussion, we sometimes presented Kruskal's algorithm, one of several known algorithms for solving this problem efficiently. This method of finding an optimal solution consists simply of repeatedly paving a shortest street which does not form a cycle of paved streets, until no further paving is required. It is interesting that the children often discovered some of the essential elements of Kruskal's algorithm and could offer arguments supporting them. (Rediscovering Kruskal's algorithm is not the point, of course.)

The natural questions that turn up in discussing this problem are rich and varied, and include such matters as, "How can you quickly tell if a proposed paving scheme meets requirement (1)?" "How can one determine if a solution can be improved?" "What is the minimum *number* of streets paved in an optimal solution?" A variety of interesting observations can (and will) be made.

In projections like this, it is not important that the teacher anticipate in advance the nuances of the problem-solving discussion that will be generated. What is important is that the children are presented with a plausible and engaging story-problem, which provides a rich field of play for commonsense mathematical reasoning. What is important for the teacher to do after the problem has been posed is to encourage and facilitate invention and discussion. (Knowing Kruskal's algorithm and other background material is not important.) When children work on problems of this sort, a rich structure of observation, argument, and solution strategies will *always* emerge.

This problem can be presented to classrooms of children aged five to six by using maps with distances marked by ticks rather than numerals, so that the total amount of paving can be figured by counting rather than by sums.

Map coloring. Maps are passed out, and it is explained that in map coloring, two countries which share a border (such as Canada and the United States) should be colored with different colors. The story concerns the poor map colorer, trying to make out a living with just a few crayons. How hard is it to tell whether two colors are enough for a given map? (There is an easy way to answer this.) How difficult is it to tell whether three colors are enough? (No one knows an easy method, and indeed the seemingly innocuous question of whether there is an easy procedure for finding the answer is equivalent to what is widely regarded as the most important open problem in computer science and the foundations of mathematics [**GJ**].)

The best solutions found for the maps under consideration (that is, the solutions using the smallest number of colors) can be displayed as attempts are made to improve these solutions. Is there a way to tell if a solution cannot be improved?

This problem is extremely rich with possible strategies, observations, and ideas. For example, one idea that often turns up is to use one color on as many countries as possible before beginning to use another color. The fact

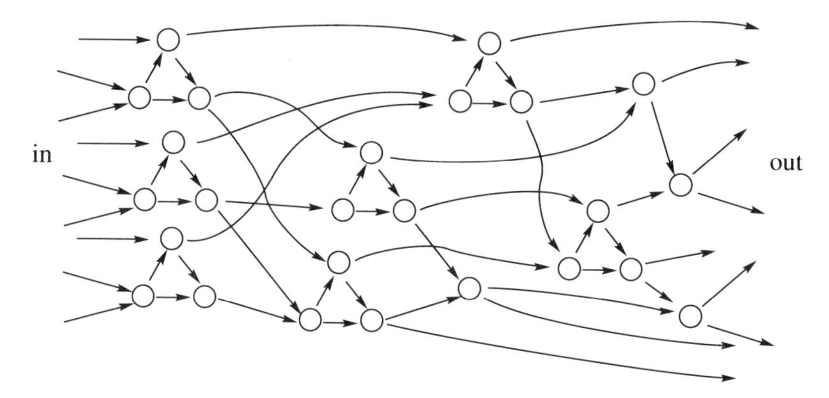

FIGURE 2. An example of a sorting network.

that four colors are always enough (the Four Color Theorem) was occasionally discussed.

Sorting networks. A sorting network, an example of which is shown in Figure 2, has n input lines and n output lines. Each *comparator node* of the network has two input lines and two output lines and functions in the following way. Regardless of how a pair of values arrive on the input lines to a comparator node, the largest value exits on the bottom output line and the smallest value exits on the top output line. The network has the property that, however a set of n values is supplied to the input lines, the values will emerge in sorted order (increasing from top to bottom) on the output lines.

Groups of children were given the project of building sorting networks from pictures in a book [**Kn**]. The networks were constructed with colored tape on a large area of lineoleum floor. (On another occasion the construction was outside in a paved area using colored chalk. A network having 10 inputs is of reasonable size.) The network was 'operated' by having ten children walk through it, carrying values (numbers or words for alphabetical ordering) on slips of paper. On following a line to a comparator node (marked by a circle) a carrier waits until another carrier arrives on the other input line to the node; they then compare their values and decide which exit lines to follow as they continue through the network.

Ice cream stands and fire stations. This problem is known in the mathematics and computer science literature as the problem of finding a *minimum dominating set* in a graph [**BM** and **GJ**]. This has many practical applications and is a classic computational problem of computer science.

The story goes like this. In order to prepare Sixtown for summer, we decided to build ice cream stands on various corners so that from any corner in the city one could reach an ice cream stand by walking at most one block. We wished to be efficient, and the problem was to find a solution using a

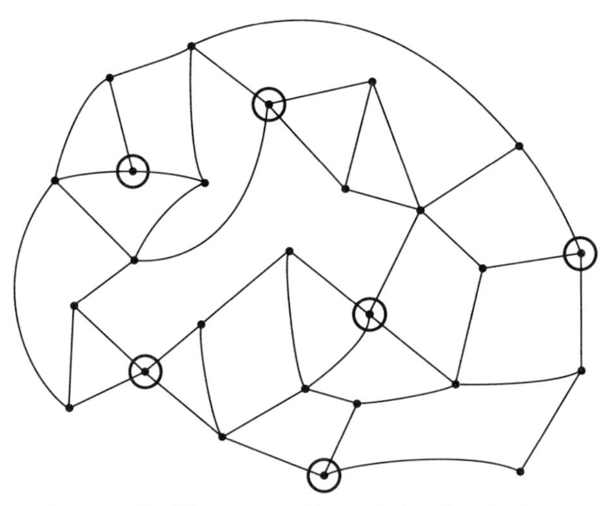

FIGURE 3. Sixtown and a minimal solution.

minimum number of stands. See Figure 3 for a map of Sixtown and a solution requiring only six stands.

On another occasion, the same problem was posed with a story concerning the placement of fire stations. It is, of course, not necessary to know what the minimum number required for a given map *is* in order to pose this problem. Many interesting observations and approaches will *always* emerge when this problem is worked on.

There are several interesting wrinkles to this problem. The author created the map of Sixtown by *beginning* with a much simpler figure for which the solution shown in Figure 3 is obvious, and then adding further 'disguising' lines. By working in this way, it is possible to create town maps for which one (privately) knows a very efficient solution (because it has been 'built in'), but for which it is often very difficult for someone else to find any solution equally efficient. This is an example of the beautiful and fundamental concept of a *one way function*, one of the conceptual building blocks of modern cryptography. On several occasions this topic was explored further. The children were charmed by the idea of creating maps which stymied their parents but for which they knew a secret solution.

There are literally dozens of graph-theoretic concepts such as dominating sets, for which stories can be invented or taken from the scientific literature, to create a rich playground for mathematical exploration and invention [**BM**, **GJ**, and **Ro**]. In the mathematical literature, concepts are often first introduced with a real or imaginary story of an application. One of the important aesthetic criteria at work in the mathematical sciences, especially computer science, is that an interesting concept is one that has an interesting story. If you can invent a mathematical problem with a good story, you have invented a problem worth exploring. There are many beautiful mathematical concepts yet to be invented by acts of storytelling.

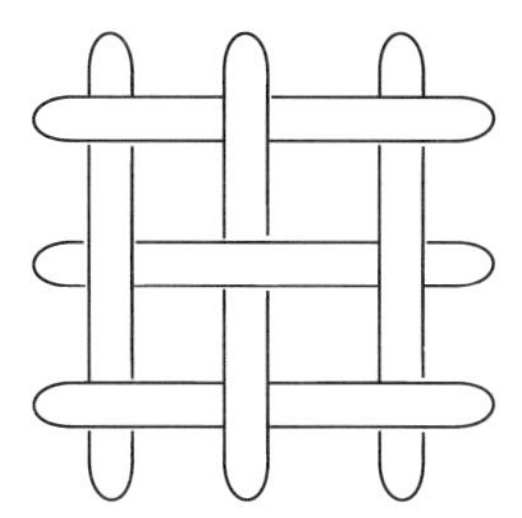

FIGURE 4. A popsicle stick exploder.

Popsicle stick exploders. This is not a classical topic, but it will serve as an example of the kind of mathematical problem which anyone could recognize or invent. In Figure 4 is shown a diagram of one possible construction of an exploder using six sticks. When the sticks are woven together as indicated, the tension from the flexing of the sticks renders the ensemble stable. If the structure is thrown with a small amount of force against a wall, it explodes with sticks flying in all directions as the flexing tension is released.

Natural questions for exploration include, "What is the minimum number of sticks with which you can construct an exploder?" "How many different exploders can be constructed with n sticks?" "Is is possible to construct arbitrarily large exploders?" It would be entirely possible for a class of nine- to ten-year-olds to investigate a topic like this and publish their findings in a mathematics journal.

There is an amusing anecdote which goes with this topic (and there are similar anecdotes being held in reserve about 'staging problems' with several of these topics). In presenting this topic to a group of children aged four to seven, the construction shown in Figure 4 was first demonstrated with the intention that the children would then experiment with constructions. The children were extremely eager, of course, as they love all things that explode. Within fifteen minutes, however, over half the class had been reduced to tears of frustration! Unanticipated in posing this problem was the amount of hand strength and dexterity required to assemble the constructions. (This problem can be alleviated by using thinner sticks.)

Knots. For all the fuss in the mathematics education literature over mystified notions such as *the concept of number* and (relevant to this topic) *spatial sense*, it is amazing that such an illuminating and beautiful geometric topic as knot theory is invariably neglected. And this while children are *playing* in the schoolyard with braids and cat's cradle! Knots, something that everyone in the world uses, have a mathematical theory that figures significantly in intellectual current events in physics, chemistry, pure mathematics, and biology **[St]**.

Knots need no special introduction, only the explanation that for the mathematics of knots, the ends of the rope are joined. This topic is obviously fun

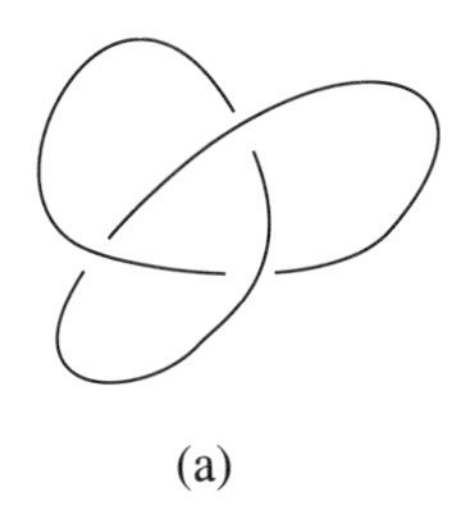

(a)

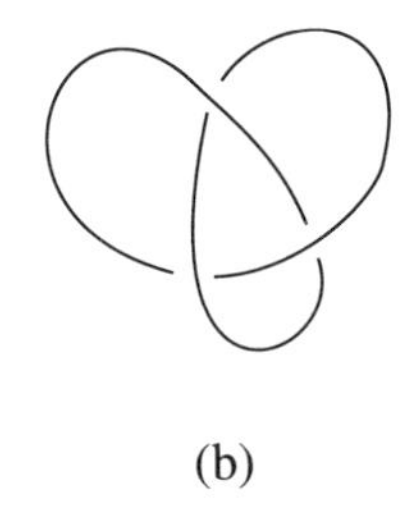

(b)

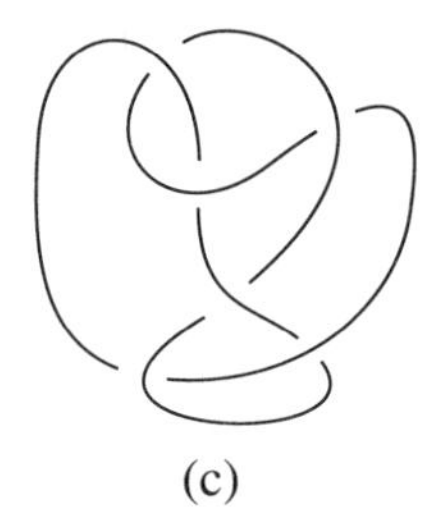

(c)

FIGURE 5. Knots.

to explore with lengths of cord and tape for joining the ends. There are a variety of intriguing questions that can be explored manipulatively. For example, is it possible to turn the knot of Figure 5(a) into the knot of Figure 5(b)? (The answer is, 'no'.) The knot depicted in Figure 5(c), however, is the *same knot* as its mirror image.

Fascinating theorems that can be explored by manipulation include the No-Unknotting Theorem **[Ga]** and the result that every knot can be put into the form of a closed circular braid. Both of these theorems have an element of the mathematical quality of surprise, and therein lies the charm of trying them out through manipulation. Apart from any theorems, there are many ways to play inventively and mathematically with knots, formulating and exploring challenging questions (some of which are presently significant open problems).

Optimal small network constructions. This is the Age of Networks, and there are many vital applications in modern technological systems of many kinds of small network designs optimizing a variety of network properties. Figure 6, for example, shows the largest known planar graphs for a few values of the parameters *maximum degree* and *diameter.* The *maximum degree* of a network is the maximum number of lines incident with any node. The *diameter* is the maximum distance (counting the number of lines to be traversed in a shortest route) between any two nodes in the network. Thus, the diameter measures the maximum number of times that a message sent through the network might need to be relayed.

It is not known whether or not there are larger networks than those shown for the indicated parameter values. This is a good example of a kind of mathematical problem on which children could actually do research nearly as well as trained mathematicians. The reason is that training is of essentially no help for this sort of problem where the combinatorial object is small. If larger networks exist, they will be found by paper, pencil, intuition, and experiment: 'no experience or background required.' There are many problems of this sort involving tradeoffs of various parameters, having applications in many different kinds of network engineering.

On the classroom visits began with a description of this research problem and an offer of a reward consisting of a lunch date and a trip to the book-

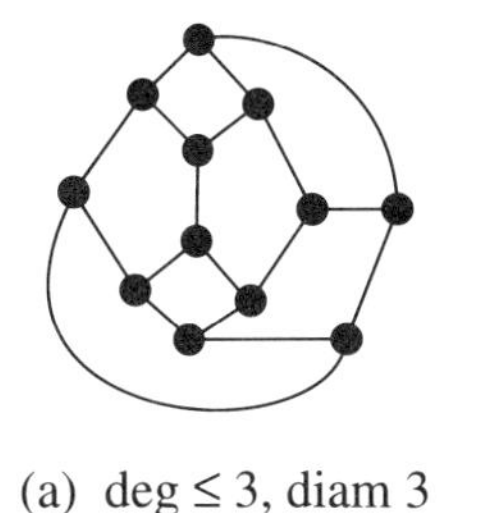

(a) deg ≤ 3, diam 3

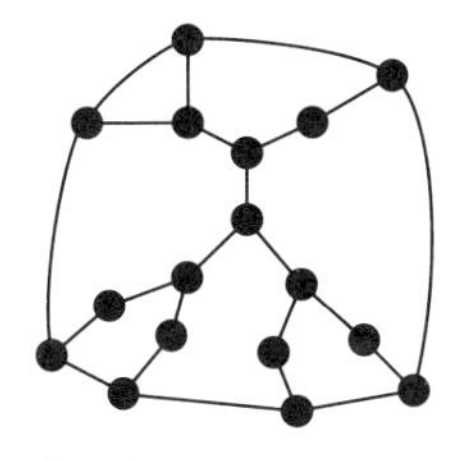

(b) deg ≤ 3, diam 4

FIGURE 6. Degree/diameter constructions.

store for anyone finding a larger construction than the largest ones presently known. A model of the network shown in Figure 6(a) was built on the floor using colored tape, and the students hopped around on this, experiencing first-hand that it did indeed have diameter 3. Note that this is an interesting and potentially valuable encounter with logical quantification, since the diameter property of the network is that *for every* pair of nodes, *there exists* a route between them of length at most 3.

The children were charmed and excited by the information that reward offers for the solution of mathematical problems is a part of the mathematics culture and by my stories about a certain famous elderly mathematician who has made many such reward offers. An impressive amount of work on this problem continued for several weeks.

Other topics. The few project areas described above hardly scratch the surface of the treasury of accessible and engaging topics in computer science and combinatorial mathematics. Sorting algorithms (sequential, parallel, randomized, on networks: there are many varieties) can provide much amusement and food for thought and opportunities for game-like physical activity [**Kn, Kr**]. The puzzle books of Smullyan are a wonderful source of mathematical and logical riddles [**Sm**]. We experimented with randomized algorithms to decide Who Pays for the Tea. We enacted the Game Show Problem (recently popularized by the radio show Car Talk and an article in Parade Magazine), keeping statistics as each student played the role of the contestant. We played Search Number and other games on graphs [**GJ**]. We planned the route of the Traveling Salesman [**GJ**], devised One-Way Street Assignments [**Ro**], Cut Stock to build the doghouse [**GJ**], and we had still hardly scratched the surface.

4. THE SEARCH FOR LEGITIMACY IN MATHEMATICS EDUCATION

This part of the story concerns explorations in the rather foreign (to this researcher) world of mathematics education, seeking some justification and sympathetic connections for the exciting and rewarding classroom experience recounted in the last section. Two impressions concerning this experience were foremost:

• The enthusiasm of the children and the teachers.

A typical conversation with a child busily trying to color a map with the minimum number of colors, would go something like:

"This is really fun!"

"Yes, this is a fun kind of mathematics, isn't it?"

"THIS is mathematics!? This is mega-mathematics!"

Several teachers stated that their picture of mathematics had been changed forever.

• The enthusiastic support of the mathematics and computer science research communities.

As an example of the latter, when the motion to stamp a committee on education to assemble the SIGACT Compendium of Theoretical Computer Science for Children was offered at the business meeting at STOC 1992 in Victoria, it was passed unanimously and enthusiastically. This is entirely consistent with the experiences of the author whenever the intellectual possibilities for children and experiences such as described in the last section have been discussed informally at research meetings.

In contrast to the encouragement received from the classroom and research communities, initial feedback from the world of mathematics education was notably *discouraging*! For example, an official of the NCTM responded to the enthusiastic tale of the classroom experiences described in the last section with the statement, "Well, I hear that *you* are having a lot of fun, but how do you know that what you are presenting to the children is *balanced*? What is your *organizer*?"

The major differences between the approach embodied in the program of classroom experiences described in the last section and what seems to be the prevailing discussion in the world of mathematics education can be summarized:

• With the exception of the Berkeley Family Math program [**Er** and **STC**], the word 'fun' and the spirit of intellectual community and excitement seems difficult to find in discussions of mathematics education and its possible reform. To a researcher, this seems to be a significant omission in a world that overwhelmingly relates to mathematics with fear and loathing while the mathematicians are having so much fun.

• One of the basic goals of the classroom experiences described in this paper is to bring to young children engaging, active, open-ended, story-full mathematics topics. The selection principle is basically 'anything goes'. Anything that can be made accessible enough to be interesting to the kids, especially if it stretches their experience of the mathematical world with objects, questions, and problems they have not yet encountered. In contrast, virtually all discussion of content curriculum in the mathematics education world is enslaved to hierarchically organized conceptual bus schedules embodying pessimistic assumptions about children's emerging abilities and interests, and inexcusably static and narrow assumptions about the nature of mathematics.[1]

• Another goal of the classroom experiences described in this paper is to bring to young children an appreciation of the frontiers of human knowledge —an appreciation of some of the mathematical questions to which no one presently knows an answer. We intend to compete with dinosaurs and outer space! Discussion of this objective appears (to the author, admittedly an outsider) to be essentially absent from the mathematics education literature.

• Nowhere in the mathematics education literature does there seem to be a discussion of the possibility of real research projects that can be actively pursued by children.

In addition to the above qualitative differences between the classroom experiences described in this paper and the main points of view articulated in the mathematics education literature, there seems to be a remarkable *timidity* even in the best of that literature. For example, consider the suggestion that an appropriate topic to present 'algorithmic thinking' to the 'prealgebra' age group would be to have them write down a detailed list of instructions for placing a long distance phone call [**Do**]. That would be about par for the excitement level all the world has come to expect of school mathematics.

Most of the mathematics education literature shows little or no awareness of the tremendous developments going on in mathematics, its modern applied branches, developments that are changing our picture of what mathematics *is*, and its role in human affairs.

In trying to understand the strangeness of the educational world and the peculiar and definitely negative role of the mathematical sciences in that realm for most people, it seemed reasonable to put forward the following hypotheses for further investigation.

A paranoid theory of mathematics education. It is sometimes offered that the importance of the traditional school mathematics curriculum is to teach children the 'discipline of thinking'. Yet it seems far more likely that the traditional curriculum serves an abusive hidden agenda contrary to the development of critical intellect and the spirit of inquiry and problem-solving. This hidden agenda may include:

(i) that the ruling social classes and authority structures do not prefer an inquiring and numerate public confident in its problem-solving ability;

(ii) that school mathematics provides excellent training in the obeying of arbitrary obscure procedures in the context of penalizing supervision,

(iii) that school mathematics provides a model for mystery cloaking the power of authority and can be effectively used to instill a sense of inferiority and self-blame on students, and

(iv) that ability in school mathematics provides a convenient rationaliza-

[1] There is a body of recent literature that we would characterize as 'good rap, bad examples' that talks about creating something resembling a research environment in the classroom, and then falls back for illustration on the usual limited and boring material [**DMN**, **NCTM**].

tion for sorting children into opportunity tracks by social class or race (as ability in Latin once was used).

What if the traditional mathematics curriculum is not really about *mathematics*, but rather, in some large measure, about *authority, power, and social rationalization*? I think the question needs to be raised. The paranoid theory, whatever its ultimate merits, has had an interesting life. At the MER workshop, the author was asked to wait in the middle of the talk while a number of people copied down the paranoid theory. Privately, several prominent mathematicians have offered the opinion that there is considerable truth in it.

5. Teaching Mathematics as a Language

Eventually, a point of view in educational theory was located that seems to be sympathetic to the kinds of classroom experiences described in §3. This perspective was found not in the mathematics education literature, but rather (surprisingly?) in the literature and community of education in the language arts. In this section we describe this point of view and how it can be applied to the teaching of mathematics—the language of science.

Recent years have seen a profound shift of perspective in the language arts education community towards a point of view that is sometimes termed *whole language.* Rather than a specific set of practices, this is a perspective on language acquisition that has classroom implications extending far beyond literacy [**AEF**].

The whole language perspective. The whole language point of view has its roots in a large body of recent research in linguistics and cognitive psychology on language acquisition [**Go1** and **GoII**]. The central fact to which this research points is that children acquire language through actually using it in a community of language users, not through practicing its separate parts until some later date when the parts are assembled and the totality is finally used. Language competency develops in a child in a way that does not depend upon instruction and drill; this fact is a central reference point in modern linguistics.

The whole language perspective based on research in linguistics and cognitive science can be summarized in its essentials as follows:

- The model of acquisition through real use (not practice exercises) is the best model for thinking about and assisting with all forms of language learning and learning in general.
- Language competency is a complex interactive system with many parts (purpose and pragmatics, syntax and semantics of cuing systems, social context, etc.) and it is not reducible to those parts.
- The development of language competencies in a child is seen as unfolding naturally and incidentally when that language is a part of the functioning of a community.

• Whole language classrooms seek to provide a richly varied and engaging environment of real language usage. The class is a community of language users, and the task of the teacher is to monitor and assist individual students in their projects, to diagnose any 'stuck points,' and to encourage the competencies presently under construction by the individual.

For lengthier descriptions of the whole language perspective on reading and writing, see [**Go1, Go2, Goll**, and **Ne**]. A few works for teachers relating the whole language perspective to mathematics education have recently appeared [**BB** and **BSS**].

A second look at our project. From the whole language perspective, the perspective embodied in the classroom visits described in this paper suddenly makes sense! The primary concern is not with a scheduled hierarchy of skills, but rather with providing a mathematically rich environment, utilizing whatever interesting material is handy.

The exercise of routine skills, such as addition, was incidental to problem-solving. For example, in the Muddy City problem, the lengths of the paved streets must be tallied.

The goal of presenting visible frontiers of knowledge can be viewed in the whole language perspective as part of constituting the classroom as a community of mathematical language users and as welcoming them to the larger community of mathematical literacy. The open-ended and exploratory nature of the mathematical topics and projects with the children made the classroom a research community. The functioning of this community while working on a topic involved a rich mix of verbal, written, social, and thinking activities.

Mathematical thinking and language. In a whole language classroom, the context for real reading and writing is often supplied by other subjects [**AEF**]. The enrichment of childrens' mathematical environment by supplying a wide-ranging experience of collective mathematical problem-solving in a classroom which functions as a mathematical research community can provide valuable opportunities for exercising and sharpening important kinds of language skills. Having an engaging playground of opportunities for the kinds of language tasks involved in articulating precise questions, presenting and justifying logical reasoning, etc., has obvious value for written and oral language mastery. In the end, a whole mathematics curriculum may strengthen a whole language curriculum.

6. The Classroom Research Community

It can be seen that the easiest way to summarize the project described in this paper is that it attempts to convey to elementary school classrooms the experience of participating in the mathematical science research community. The idea that the classroom should function as a literate community of readers and writers is central to the whole language approach to the teaching of language skills. The project described in this paper is thus fundamentally in tune with this outlook.

What aspects of the mathematical science research community experience are portable to elementary school classrooms? Some vital parts of that experience most definitely *can* be brought to the elementary grades.

Playfulness. It is easy for a practicing scientist to share this part of the experience of participating in the research community with children in the earliest grades, because young children still know vividly how to play and have some natural solidarity as a research community.

Kurt Vonnegut pointed to this commonality in an amusing way when he remarked, "If you are going to teach, you should either teach graduate school or fourth grade... And if you can't explain it to fourth graders, you probably don't know what you're talking about."

Playfulness is related to abstraction and modeling. Presented with a map of the Muddy City, young children are quite comfortable with regarding the dots as representing houses, etc. (After all, they just finished asking their parents to regard an odd bit of stick or a paper cut-out as a laser gun.) College students are more likely to complain about a perceived lack of 'realism' in the model.

Playfulness is often deeper than it appears. For example, from a 'serious' perspective, the problem of the map-colorer eking a living with a few crayons may seem to be a fairy tale and a silly waste of time, compared to 'real' and 'practical' school mathematics such as the mechanics of long division. This fairy tale problem is both mathematically profound and has many important industrial applications. Do not forget Einstein's famous advice about what physics texts are most important to read: "Fairy tales, and more fairy tales." It is the sense that mathematics permeates the world in this way, that informs the aesthetic in the research community that loves a concept that has (or makes) a good story.

Asking questions. Of course, the research community is always asking and trying to answer questions. In the mathematics research community, a good deal of recognition can come from just asking a good question, quite apart from being able to answer it. Question-asking can be made an important part of the classroom mathematics community; there is essentially no place for it in the traditional school mathematics curriculum with its hopelessly narrow and petrified view of mathematics.

Associated naturally with the importance of question-asking, is that it is all right not to know the answers. This makes for a radical shift in the relationship between the teacher and the class in the contrasting settings of traditional and whole mathematics classrooms. To participate in the latter, teachers must see that creating an interesting problem-solving environment where children can ask questions for which the teacher does not know the answer is *positively* a good thing. It is *not* necessary, or even desirable, to know the answer (or the 'background') before posing the story-problem.

A complex relationship to truth. In the traditional school mathematics classroom, truth plays a strikingly simple role, in stark contrast to both everyday problem-solving of any kind and to the complex experience of truth in the mathematics research community. Mathematical statements, as they occur in the research community, are richly colored in a variety of important ways that tend not to be well appreciated presently outside of that community. A statement may be intuitively clear and have an easy proof, or it may be strongly intuitive (for example, that the first two knots in Figure 5 are not equivalent) but have only a difficult proof. A statement may be intuitively clear, but have no known proof. And it can sometimes turn out that intuition is wrong! Good mathematics requires both unrelenting skepticism and wild imagination.

Much could be said about the humanizing value of a rich experience of truth. This experience can be encouraged in classrooms where children formulate, discuss, and explore conjectures. These may concern such questions as, "How big is the moon?" or "How many berries are in the berry patch?" There are also beautiful unsolved mathematical problems that can be shared with children. One that we shared in elementary classrooms is illustrated in Figure 7.

The conjecture is that if you create any sequence of letters (using an alphabet of any size) where each letter appears three times, it will always be possible to select one copy of each letter in such a way that no two selected copies are adjacent in the sequence. Figure 7 shows an example of such a sequence, where the alphabet is $\{A, \ldots, H\}$ (eight letters). Since each letter appears three times in the sequence, the sequence has length 24. The eight arrows select one of each of the letters. Note that no two of the selected letters are adjacent in the sequence, i.e., no two of the arrows are immediately next to each other.

This conjecture remains unproven despite strenuous efforts by a number of mathematicians to settle it. There are many such beautiful and sometimes famous conjectures concerning combinatorial patterns and relationships that are accessible to young children and that can be manipulatively explored.

Communication. The tremendous importance of communication, and the fact that the bulk of one's time as a scientist is not spent in discovering things

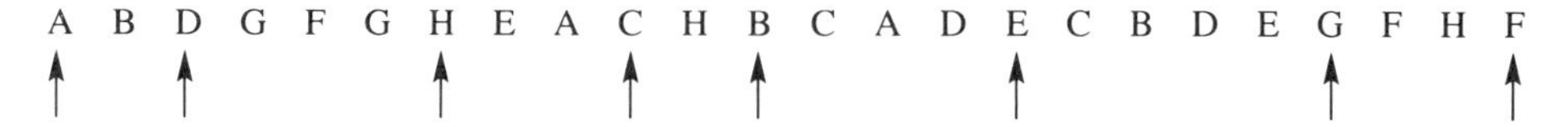

FIGURE 7. An open problem of discrete mathematics.

but rather in communicating those discoveries, tends to be not widely appreciated outside of the science research communities. The traditional school mathematics classroom, with the emphasis on silent individual seatwork with very little writing, could hardly be more different in character. In the mathematics research community, many different modes of communication have a vital role, including electronic mail, formal, informal, and very informal verbal communication, and carefully written archives. The classroom mathematics community can also be conducted in such a way that a variety of communication modes have an important role.

Participation in larger research communities. As mentioned above, some mathematical problems in combinatorics could actually be investigated by elementary school children, although one might expect these opportunities to be somewhat limited. This is one way in which the classroom research community can be connected to larger research communities.

Another possible connection that a classroom can make is to communicate with mathematicians and computer scientists in the community, asking them for answers to questions, or inviting them to visit.[2]

The same could be done with any of the sciences. The mathematical sciences have a slight advantage in providing a model for classroom research communities in that their laboratories are so portable!

7. An Unexpected Windfall

It may be of interest to professional mathematicians to hear that the classroom experiences described in this paper led to some interesting research problems. Here are a few stories of such interaction.

• There were staging difficulties with presenting the zero-knowledge protocol for small dominating sets with seven-year-olds (one of our more ambitious projects). Blum's algorithm requires quite a few envelopes, and these were, to put it bluntly, *incompetently numbered* (with transposed and illegible backwards numerals, etc.) by one of our assistants. The protocol is relatively sensitive to having a correctly labeled set of envelopes! But this raised an interesting question. How may envelopes do you actually need for a round that yields certainty at least $1/2$? A research paper on this very topic appeared at about the same time.

• The sorting network project was extremely popular and the kids requested a repeat. On the appointed day, I was very short of time and of my copy of Knuth. I thought I might quickly work out from scratch a sorting network for eight inputs, but it was quite difficult! With the pressure on, it occurred to me that a certain randomized construction method might work. We tried that, and it did work. With hindsight, it was easy to make a heuris-

[2]A touching example of this was the material given by one young student to a colleague to take to the MER workshop, "to show to the mathematicians and see what they think."

tic argument for the success of the method, but a theorem would be more difficult. I mentioned the problem to a colleague shortly afterwards, who subsequently became involved in writing a paper on this subject [**LP**].

• Last spring, a colleague, Jan Kratochvil, visited for several months from Prague, and graciously came along on several of my school visits (which was easy, since he rented a room at the Apple Blossom Family School). After several of these sessions, we spent some hours trying to solve problems that 'turned up' on these adventures. For example, children quickly grasp the idea of a proper coloring of a map. A natural introduction for a child to possible strategies for minimizing the number of colors might involve taking turns coloring regions. The Four Color Theorem assures, if you are playing perfectly (and by yourself), that four colors are enough—but what if you alternate turns with 'incompetent help' who make moves that are at least legal? An upper bound of 33 (probably not tight) was established by Kierstead and Trotter in 1992 [**KT**].

• One of the projects with one-way functions involved creating graphs for which one knows a small dominating set, which can be very difficult (apparently!) for anyone else to find or match in size. But a theorem that would make this explicit must grapple with a thorny conundrum of complexity theory. Although, for example, there are more-than-polynomially many hard instances of an NP-complete problem such as Dominating Set (unless $P = NP$), it may be that the hard instances are still relatively rare, or difficult to generate. One wants a theorem (so it turns out) that P-sampleable distributions are hard for average case fixed-parameter search complexity, an interesting variation on a hot topic of current research in theoretical computer science.

In retrospect, why would not one expect this sort of valuable feedback from the experience of explaining things to children? It would seem to be appropriate if contact between graduate students (who are finally emerging into the realm where they can *play again* at learning) and small children (who are living in world of play) to share and explain live science were a regular and normal part of our research lifestyle and training. We should make a loop of shared experience like this, and gradually draw it in, until all of schooling becomes language-learning play, connected to real, current live projects.

Acknowledgments. Thanks especially to Marfa Levine, without whose persistent pursuit of an educational dream this adventure would perhaps not have materialized, and to Nancy Casey for inspiration and guidance in the ideas and literature of the whole language outlook on education. Thanks to Carole Moore and Prudy Heimsch for making their classrooms and experience available to us.

References

[**AEF**] B. Altwerger, C. Edelsky, and B. M. Flores, *Whole language: What's new?* The Reading Teacher, November (1987) 144–154.

[BB] A. Baker and J. Baker, *Mathematics in process*, Heinemann, Portsmouth, NH, 1990.

[BSS] D. Baker, C. Semple, and T. Stead, *How big is the moon?* Heinemann, Portsmouth, NH, 1990.

[BM] J. A. Bondy and U. S. R. Murty, *Graph theory with applications*, American Elsevier, New York, NY, 1976.

[CLR] T. H. Cormen, C. E. Leiserson, and R. L. Rivest, *Introduction to algorithms*, MIT Press, Cambridge, MA, 1990.

[Do] J. A. Dossey, *Discrete mathematics: The math for our time.* In M. J. Kenney and C. R. Hirsch (eds.), Discrete Mathematics Across the Curriculum K–12, National Council of Teachers of Mathematics, Reston, VA, 1991, pp. 1–9.

[DMN] R. B. Davis, C. A. Maher, and N. Noddings (eds.), *Constructivist views on the teaching and learning of mathematics*, National Council of Teachers of Mathematics, Reston, VA, 1990.

[Er] T. Erickson, *Off and running*, Lawrence Hall of Science, Berkeley, CA, 1986.

[Ga] M. Gardiner, *Knotted doughnuts and other mathematical amusements*, W. H. Freeman, San Francisco, CA 1982.

[Go1] K. Goodman, *Reading: A psycholinguistic guessing game*, in H. Singer and R. Ruddell (eds.), Theoretical Models and Processes of Reading, International Reading Association, Newark, NJ, 1976.

[Go2] ——, *What's whole in whole language*, Heinemann Educational Books, Portsmouth, NH, 1986.

[Goll] F. Gollasch (ed.), *Language and literacy: The selected writings of Kenneth S. Goodman, Volumes 1 and 2*, Routledge and Kegan Paul, London, UK, 1982.

[GJ] M. R. Garey and D. S. Johnson, *Computers and intractability: A guide to the theory of NP-completeness*, W. H. Freeman, San Francisco, CA, 1979.

[Kn] D. E. Knuth, *The art of computer programming, Vol. 3, sorting and searching*, Addison-Wesley, Reading, MA, 1973.

[Kr] L. Kronsjo, *Computational complexity of sequential and parallel algorithms*, Wiley, 1985.

[KT] H. A. Kierstead and W. T. Trotter, *Planar graph coloring with an uncooperative partner*, manuscript, April, 1992.

[LP] T. Leighton and C. G. Plaxton, *A (fairly) simple circuit that (usually) sorts*, Proc. 31st Annual Symposium on Foundations of Computer Science (1990), 264–274.

[NCTM] *Curriculum and evaluation standards for school mathematics*, National Council of Teachers of Mathematics, Reston, VA, 1989.

[Ne] J. M. Newman (ed.), *Whole Language Theory in Use*, Heinemann Educational Books, Portsmouth, NH, 1985.

[Ro] F. S. Roberts, *Discrete mathematical models with applications to social, biological and environmental problems*, Prentice-Hall, Englewood Cliffs, NJ, 1976.

[Sch] A. H. Schoenfeld, *Problem solving in context(s)*, in R. I. Charles and E. A. Silver (eds.), Research Agenda for Mathematics Education, Volume 3, The Teaching and Assessing of Mathematical Problem Solving, National Council of Teachers of Mathematics, Reston, VA, 1989.

[Sm] R. Smullyan, *Forever Undecided*, Knopf, New York, NY, 1987.

[St] I. Stewart, *The problems of mathematics*, Oxford University Press, Oxford, UK, 1987.

[STC] J. K. Stenmark, V. Thompsen and R. Cossey, *Family Math*, Lawrence Hall of Science, Univ. of California Press, Berkeley, CA, 1986.

Computer Science Department, University of Victoria, Victoria, B.C., Canada V8W 2Y2
E-mail address: mfellows@csr.uvic.ca

CBMS Issues in Mathematics Education
Volume 3, 1993

Promoting Mathematics Learning Among Minority Students: Critical Issues for Program Planners and Mathematics Educators

CHARLES H. ROBERTS

In recent years, increased attention has been directed at efforts to promote mathematics learning among students from certain groups in society. In particular, a group is identified as underachieving in mathematics because its performance rankings are below national norms or because the group's representation in the study of mathematics diminishes as the group moves through the educational system. Experience has shown that even among the best students from certain ethnic minority groups there is a need for special, and often extensive, attention in order for them to reach their potential in the study of mathematics at both the pre-college and post-secondary levels. At the other extreme, many students need both extensive and intrusive intervention before they can begin to realize any measure of success and lay a foundation upon which to build for improved performance. Thus any effort to address the problem of minority students' performance levels in mathematics needs to have as part of its basis a strong consideration for the many related issues and concerns associated with the problem. I claim that five elements are crucial for success:

(1) Program planners must gain a thorough understanding of the affected students and their problems;

(2) Based on this understanding, planners need to decide on the appropriate philosophical foundation upon which to conduct the effort;

(3) Certain aggressive measures must be taken to successfully negotiate within the system on behalf of the students;

(4) Students must be induced to incorporate specific methods and strategies into their daily routines in order to enhance their learning skills and reach high levels of performance; and

(5) The effort must be thoroughly evaluated.

©1993 American Mathematical Society
1047-398X/93 $1.00 + $.25 per page

Social Basis

The need to understand students and their problems is especially crucial with respect to many ethnic minority students since much of the explanation for their lower performance levels as compared to standardized norms is related to cultural differences, historical conditions, and personal experiences as minorities. Fortunately the increased attention to ethnic minority students has resulted in the availability of much more information on cultural differences. However, there is much to be learned, especially regarding the cultural dynamics which effect minority students' ability to function successfully in the traditional educational setting in this country.

Due to cultural differences, which in large part are due to differences in experiences in society, minority students often possess a different perception of reality than that of the majority culture. For example, upon entering a new educational institution, minority students might form opinions about the institution which are substantially different from their majority counterparts because of their perception of a lack of concern for them and their cultural heritage (or even perhaps a perception of hostility toward them) on the part of faculty and staff members of the institution with whom they might have initial interactions. This perception could, in turn, negatively influence their disposition toward seeking out or developing a rapport with those whose responsibility it is to guide and assist students, despite the fact that the students' growth and development in the system might depend on it. Thus, without a prior understanding of this phenomenon, neither the students themselves nor the institutional personnel will be in a position to anticipate students' concerns and begin to take appropriate steps to address them.

Perhaps a more thorough understanding of the nature of social and cultural oppression among certain minority groups and its impact on students' ability to function successfully in our educational institutions would be of great assistance. That members of oppressed groups suffer certain long-term performance inhibiting consequences as a direct result of the oppression itself is well documented. For example, cultural domination by one group over another in society might retard creative expression in academic enterprises of many individuals of the affected group. Thus, cultural domination can have a negative impact on the dominated group's academic performance. It is incumbent upon us as mathematicians and/or educators to carefully explore this issue and determine appropriate action to counteract the negative effects, as well as to take measures to eliminate inhibiting factors.

Establishing Beliefs

In educational endeavors, the establishment of a sound philosophical foundation can often mean the difference between success and failure in reaching set goals. For example, what beliefs do we hold about students, their capabilities, and the likelihood that we can positively affect their performance levels?

Can students who possess serious deficiencies in learning mathematics overcome them and become good students over time? Can we assist students in overcoming the debilitating effects of centuries of social and cultural oppression and empower them academically? To what extent can students improve their performance through their own behaviors? Do we have the capacity to significantly alter students' dispositions toward learning mathematics? To what extend should we intrude on students' personal lives in the interest of assisting them? And how comprehensive and/or extensive must the effort be in order to make a measurable and considerable difference?

It is apparent that if we are to induce a high level of performance in mathematics and science among minority students, we must firmly believe that they can achieve at the desired level, and that we will be able to find ways of providing opportunities for them to do so. With this belief, we would then seek out ways of assisting students in not only overcoming existing barriers but in finding the means to direct their own academic development within the system as well, provided, of course, that the requisite resources were made available. It is only after this effort has become entrenched that we will ever be able to know the extent to which students are able to take responsibility for their own learning.

Intrusive Involvement

In attempting to assist students to develop prerequisite skills and dispositions needed to succeed in mathematics, we must ask ourselves to what extent we are willing to get personally involved with students and to how expansive an effort we are willing to commit ourselves. Due to the depth and pervasiveness of the problem at hand, we are likely to find that, in order to effectively address it, a dynamically intrusive and encompassing effort must be planned, the kind that is perhaps rare in educational endeavors today. An effort of the magnitude which might be needed would likely require the involvement of educators and/or professionals from other disciplines, such as psychology and sociology, as well. These and other similar issues must be aggressively tackled throughout the process of planning for and operating any intervention effort which would thoroughly address the problem at hand.

Many crucial decisions related to incorporating a consideration for minority students' concerns into the academic processes and integrating program efforts into the academic mainstream must often be made as well. For example, if the affected students attend school in a school system in which they are in a minority, to what extent should we single them out for special assistance in order to address their particular needs? What about after-school sessions? What special efforts to reach out to them should be made in the classroom? Should all teachers, parents, and perhaps other selected figures be expected to become substantially familiar with the issues being outlined here? Or should only certain designated teachers and/or administrators become the 'experts' on these issues? Finding agreeable answers to these questions might prove quite difficult, but they must be wrestled with nonetheless.

Cultural Relevance

Regarding the curriculum, many of us have come to strongly contend that unless a determined commitment to including a concern for cultural relevance is made, we will continue to fail to adequately prepare minority students in the study of mathematics, science, and other disciplines. Maybe we should include word problems which are more directly related to the experiences and backgrounds of these students. For example, when teaching percents, problems which explore population shifts over certain periods of history for certain minority groups might prove interesting to students while informative for everyone. Or perhaps more classroom discussion should be directed at students' lives and more attention paid to helping them make important connections between their world and the world of mathematics. Examples which have a local flavor, such as using street patterns in the teaching of geometry and/or trigonometry, would seem to promote this effort.

Confronting Barriers

At the post-secondary level, some questions, such as whether it is feasible to provide out-of-classroom support for minority students, might not be as problematic as they could be at the pre-college level, while other concerns, such as gaining access to the desired population of students, might prove more problematic. Attempting to institutionalize initiatives might also prove problematic, depending on the level of cooperation and support received from administrators and on the in-place structure of the academic system. On the other hand, the need to make behavioral demands on students, e.g., required study practices, in order to induce them to enhance their own academic development might be hindered more by the lack of student-staff rapport than by institutional structure, though that too could play a part. Finally, a crucial concern of any effort to assist minority students on majority campuses is that of confronting the manifestations of cultural conflict often experienced by many of these students, e.g., the conflict brought on by the need to maintain a subjective world view while attempting to adhere to the demand for objectivity in the study of mathematics and science. For students who plan to continue to study mathematics and/or science, this concern could prove especially important because often these conflicts must be resolved before the students are able to tackle the many academic challenges inherent in attempting to acquire the learning skills that are needed.

Inducing Behaviors

As the above issues and concerns are being addressed and the needed intervention measures have been taken, programmers must decide what students should do in order to develop the required cognitive skills. Several examples exist to demonstrate minority students' capabilities in studying mathematics and science. However, experience has shown that determining the most effective methods of providing the 'appropriate opportunity' for a given set of students can prove especially difficult. Furthermore, ensuring that students follow through and incorporate prescribed strategies into their daily routines is also problematic. Thus program planners and facilitators alike must continually seek out the methods and strategies which prove effective in getting results among students and aggressively determine ways of prompting their students to fully adopt them.

Process Evaluation

Evaluation, the final element for success, is perhaps the most crucial. For without a thorough evaluation plan we cannot begin to understand what we have accomplished, whether intended or unintended. Figuratively speaking, we remain in the infant or perhaps toddler stage of knowing what needs to be known about why minority group students as a whole have neither performed nor progressed as well in the study of mathematics as we would like. It is important that the evaluation process includes a provision for assessing the students' perceptions, opinions, and actual experiences as they make the important transition from being disadvantaged underachievers to high achievers in mathematics and science. Furthermore, it is important to delineate and characterize the factors and activities which substantially contributed to students' improved performance (e.g., student-staff dialogue, peer group interactions, etc.), and to document the extent and nature of their impact. Of course, keeping in mind at all times the goal of promoting excellence in performance, the objective measurement of students' performance relative to the larger population in society should anchor the entire evaluation process. The information and results emerging from implementing the evaluation steps above are likely to apply to the general mathematics education community. All of us would then be the beneficiaries.

Department of Mathematics, Michigan State University, East Lansing, Michigan, 48824

CBMS Issues in Mathematics Education
Volume 3, 1993

Systematic Reform: Curricula, Context, Culture, and Environment

HARVEY B. KEYNES

A. INTRODUCTION

The major goal of this conference is the substantive reform of pre-service teacher preparation. I have been asked to address the issues of the roles of mathematics curricula and teaching methods in this reform process. In my earlier and more naive days in dealing with these topics, I would have rather glibly regarded this talk as a 'piece of cake'. After all, over the past several years, there have been a large number of important curricular ideas and directions developed, and only the most cynical person would doubt that they will play a very important role in the mathematics curricula of the future. I probably would have also downplayed the role of pedagogy and teaching in reforming mathematics instruction in K–16 education. Fortunately, the past ten years in working in mathematics education and the reform movement have made me much more humble about the real issues of reform in mathematics and perhaps even a bit wiser. I would like to share with you my 'reformed' view of the real issues in changing mathematics and science curricula in the K–16 systems. Since Nebraska is one of the first states to have obtained a major five-year Statewide Systemic Initiative grant from the NSF, I am sure that several of you have addressed these same concerns and probably reached similar conclusions. This conference provides an excellent opportunity for me to go 'public' with this perspective and widen the discussion of this viewpoint of reform.

First and foremost, I want to make it clear that I do believe the content and curricula of mathematics and sciences are the keystones of any real reforms in teaching and learning. Without meaningful changes in these aspects, any other changes will probably go nowhere. However, there are other critical issues which, if not properly addressed, could minimize at best and most likely sink even the most forward-looking curricula. These issues are:

- the context of the mathematics and science curricula;

©1993 American Mathematical Society
1047-398X/93 $1.00 + $.25 per page

- the teaching and learning environment; and
- the culture and rewards of the education system.

Without significant attention to these areas, the current reform movement in mathematics could easily peak and ultimately have only a very minor influence on mathematics instruction in this country. Given the international marketplace and our need to match world standards, the United States can ill afford to lose the current momentum and have to wait another 25 years for the next reform movement.

To emphasize the importance of these noncontent issues, I will first address some of the critical aspects of these points. Then we will turn to curricular and content issues and reform ideas which will likely play an important role in the future.

B. Context, Culture, and Environment

1. Perspectives.

> *If you do what you always have done, you will get to where you always get.*
> *All changes have about a three-year life cycle. If you are patient, it too shall pass.*

The history of reform movements is realistically summarized in these sayings. We tinker with the edges and even create pockets of real excitement. Certain individuals are changed, and some second- and third-order effects on the system may be felt. But overall, the system usually functions in the same way afterwards as it did before. Perhaps in the past, the need for change was not really that important. Few mathematicians and scientists hold to that opinion today.

From many perspectives, the education system is really in need of major reform. A recent study by the National Association of Manufacturers (NAM) reported the average manufacturer rejects *five out of every six* job candidates because of a lack of basic skills. As one might expect, the major problems were poor reading and writing skills and poor mathematics capabilities. These problems persist with the existing work force. More than half of the reporting companies indicate major skill deficits among employees in basic mathematics and problem solving. As a result, about 40% of the companies reported serious problems upgrading technology and increasing productivity. Worse yet, about 30% were unable to simply reorganize work activities or upgrade product quality because workers could not learn new skills or adapt to new jobs. These surely are chilling statistics. Indeed, the American 'job problem' is no longer one of too few jobs, but of too few skilled workers for existing jobs.

Where does one begin to attack these problems? It is clear that reform must reach a larger part of the population than ever before. The 1990 National

Assessment of Educational Progress (NAEP) noted that only 16% of American 13-year-olds could compute with decimals, fractions, and percentages, recognize geometry figures, and solve simple equations. By twelfth grade, only about 46% of those still remaining in school have achieved this level. These low skill levels certainly contribute to our workforce problems. But to reach more students, it is not simply a matter of a more demanding curricula. It is also very important to improve the contextual approaches in which mathematics and science are taught. To reach more students and reverse the skill gap, even the most innovative and enlightened mathematics curriculum will need to be presented in different ways.

2. Context. It has always been good educational practice in elementary mathematics instruction to use a variety of ways to teach basic concepts. This recognizes that different students learn by different ways. As we move up the curriculum, there has traditionally been less emphasis on different approaches. The NCTM *Standards* calls for more use of real world applications, statistical and graphical data, and many forms of technology in mathematics instruction. Yet the latest evidence indicates that while the schools do succeed in teaching the lower-level mathematics skills, they have had very little success in teaching higher-order skills. We need to consider alternate approaches if we ever hope to reach more students with this level of knowledge.

In a very provocative talk entitled *Will everybody ever count?*, Lynn Steen, the chief architect of the widely read National Research Council report, *Everybody counts: A report to the nation on the future of mathematics education*, examines the question of progress toward achieving this goal. In this report, the benchmark is widespread understanding at the NAEP 350 level, an understanding of elementary algebra, geometry, statistics, and probability. Moreover, the strategy is the NCTM goal of a common core curriculum, differentiated by approach and depth, but not by objectives. Steen concludes that we are *not* very likely to reach this goal. He states [**11**, p. 5] "If by 'count' we mean those parts of mathematics that really count in our global society, then not everybody will count now or in the future." The suggested alternative is to allow variation based on numerous sources: that is, to teach and assess mathematics in context. This permits maintaining high expectations for all in a common core curriculum while allowing success to be measured in smaller increments than the standard measures like NAEP. The intent would be that [**11**, p. 6] "... all schools and all children make consistent, measurable progress towards the goal of mathematical power for all."

A good subject to examine these ideas is elementary algebra. As a keystone topic to most advanced mathematics coursework, this topic has been recently revised by many authors in many directions. Yet even the most reformed approaches to algebra rely heavily on traditional computation and algebraic manipulation as the major part of the curriculum. To succeed in algebra essentially means to succeed in standard computations and manipulations, even if the problems are put in more realistic settings. However, there is a great deal of evidence which indicates that for most students, algebraic computational skills will remain a major obstacle, even with the supplemental help provided by appropriate technology. We need a broader definition of learning and success in algebra to help engage more students, indeed virtually all students, in studying and appreciating some of the important algebraic concepts, such as variable, function, linear dependence, and inequalities. Some approaches may primarily involve only pictures, data, and simple manipulations to obtain an elementary understanding of a few ideas. But the result would be that nearly all students would have an opportunity to learn some aspects of algebra, and the system would recognize this progress.

There can be concern that by allowing for context, we downgrade standards or soften a subject. This is not the intent of variation. Great variation can and should take place in the same classroom, to encourage reasonable progress for all students. Indeed, in programs for more talented students, the contextual overview allows for deeper conceptual and computational approaches which will ignore more routine and transparent ideas. Moreover, different students will concentrate on different aspects of a topic, depending on their individual approach. This has been the principle on which the University of Minnesota Talented Youth Mathematics Program developed its algebra curriculum. Also, success with interdisciplinary mathematics/science curricula will certainly require a contextual approach. There is a real difference in the mathematics of the science classroom and the traditional classroom approaches to mathematics. Allowing variation in mathematical background would allow more students to learn and appreciate the connections between mathematics and science.

3. The teaching and learning environment.

> *African Proverb: It takes an entire village to teach one child.*
> *The greatest strength and greatest weakness of American education is local control.*
> *A pig does not get any fatter by weighing it more often.*

No matter what we do with curriculum, the real litmus test for success is what happens in the classroom. While we can be quite sure that a teacher with an inadequate background and understanding of mathematics will not teach it well, we also know that many teaching situations can also discourage even the brightest and most capable teacher with an outstanding curriculum. Experiences in a major teacher enhancement project at the University of Minnesota showed that innovative curricular approaches by highly capable teachers never reached the classroom because of school/district lack of commitment and support. Curriculum innovation will only succeed when the classroom and school environments encourage and nurture it.

The current environments in most schools are simply not supportive of effective teaching and learning. While Jonathan Kozol describes the dismal situation in the urban schools with the term 'savage inequality', many rural schools suffer in some of the same ways. Even the affluent suburban schools have a deteriorating environment due to recent budget cuts. A 1991 United States Department of Education study [**14**] surveyed public school teachers who left in 1988–89 due to dissatisfaction with teaching as a career. While 7.3% left due to salary, 26.4%—3.5 times as many teachers—quit because of inadequate support of the schools/administration! Although student comparisons with Japan and Taiwan are subject to debate, virtually everyone agrees that the professional structure of the teacher's day and the general teaching environment in Japan is superior to the American classroom.

For those of us who really want the curricula of the future to come into the classroom, we need to be willing to try and support some very different teaching environments. They must focus squarely on the teacher and provide extensive support, while at the same time expecting innovation and improvement. The NSF-supported Teaching Integrated Mathematics and Science (TIMS) program, a new and innovative K–6 mathematics/sciences curriculum, states in its current newsletter [**13**, p. 5]: "Our experience with TIMS has demonstrated that a qualitatively and quantitatively different effort is needed to truly revitalize a math/science program. For reforms to take root, an overall environment must be established within the school that truly supports innovation." The NSF State Systemic Initatives grants provide a real opportunity to change the rules of the game and try to create significant, long-term modifications to the teaching environment. We should agressively pursue these opportunities.

Turning to the learning environment, it is easy to regard the task as extremely complex and nearly hopeless to change. There are many aspects involving family, culture, health, and socioeconomic status that we in the mathematics classrooms cannot really change. Yet there are aspects of our classroom conduct and our expectations for students which can have a profound effect on a few and at least set improved standards for the others. The NCTM *Standards* provides a good overview on this issue. I would like to concentrate here on two of these aspects: the roles of family/community support and of student effort.

Although we frequently pay lip service to involving the family in education, the reality is that most families are not significantly involved in their child's education. Many feel that the schools are hostile environments. In mathematics, lack of knowledge of the subject discourages parental involvement. Yet, much evidence shows that the support of a parent or community leader can have a profound influence on student interest and output. In the University of Minnesota Talented Youth Mathematics Program, it was found [**3**] that a major key for success is the active support of a parent or other adult. This support does not require knowledge of the subjects, but simply involves providing a decent study environment, seeking assistance for the students when they are struggling, and letting their student know that participation in the program is valued and encouraged. As the program has successfully increased its population of females and students of color, this aspect has been found to be even more important. This parental support is heavily solicited by the program, and its impact described over and over again to the families. It is worth every bit of the effort. The teachers and schools need to seek similar support from the parents of all students.

We also need to honestly address the role of effort in learning mathematics. As we reform our curriculum and replace drill-and-compute skills with more challenging materials, we all know that it will take as much, if not more, effort for students to learn the subject. This effort must be expected of all students in order to maintain high standards and opportunities for improvement of mathematical skills. Whatever the level of mathematics, we need to expect an honest effort from all students, and base measures of progress on this level of effort.

While delivering the message about student effort may be a difficult job, the task is worth pursuing. We are all aware of the enormous efforts and sacrifices that students and families are willing to put into athletics. Students *can* find more time for mathematics if they value it. The Talented Youth Mathematics Program routinely demands nine to ten hours per week of concentrated homework from its students. This is in contrast to a 1991 study [**16**] of high school high achievers performed by Who's Who Among American High School Students in which 55.4% of high-achieving high school juniors and seniors studied seven hours or less per week for *all* subjects, and only 20.5% studied 11 hours or more. The program has found that this effort plays as important a role as native intelligence and actual coursework in the striking success of its graduates in demanding mathematics, science, and engineering programs at the best universities. Finally, successful Asian-American students and other immigrants nearly always attribute their success to their efforts and work ethics rather than innate brilliance.

Convincing students to put more efforts into schoolwork will mean going against many social and even cultural norms which influence their lives. It will certainly require a significant family or community effort. Using the experiences of the Talented Youth Mathematics Program, programs to educate parents or other adults who are significant in the student's life about work expectations and educational advocacy are important. Involving these adults in activities such as Family Math, which link them to the student's mathematics program, may also help. All of the Treisman programs for minority students have demonstrated the effectiveness of community group study in a socialized and cooperative setting. They provide quality study time as well as opportunities for support and cooperation. Widespread implementation of these workshops in community settings could affect the working habits of many students. These workshops may also help to start to create cultures of mathematical interests and kinship among interested groups of students. At the least, they will provide opportunities for students to focus on schoolwork in a positive and productive manner.

In summary, we need to seriously engage the students and their families in the learning of mathematics, and foster discipline, determination, and desire in all students. Zalman Usiskin, the director of the University of Chicago School of Mathematics Project, recently addressed these issues in a paper delivered at a national UCSMP conference. He states [**15**, p. 11]:

"We cannot attain the significant improvement needed in student performance unless the students themselves change. Students who are behind should not be given a weak curriculum—they need to work harder to catch up—but they will never catch up if they do not do their homework, because those ahead of them got there by working and will continue to work. This must be made clear. Changes in student behavior are necessary for the disciplined environment conducive to learning that the President[1] has called for. And these changes will take more than your individual efforts. They require the entire school community—parents, teachers, administrators, business and government leaders—working together. It is a difficult job, and in many places it will take years to achieve, but it is worth striving for."

4. The culture and rewards of the educational system.

> *Most mathematicians spend their lifetime trying to win the grudging admiration of a few close mathematical friends.*
> *We educate drivers and train teachers.*
> *The real choices of teachers in the current situation: quit and leave, or quit and stay.*

This topic probably strikes close to home for all of us in some very emotional ways. At the college and university levels, there is a renewed debate emerging on the roles of research, teaching, scholarship, and service, and the role of the reward structure in setting these priorities. At the school level, there continues to be concerns and discussions about how teachers are educated and the role of the discipline subjects, how teacher subject knowledge will limit curriculum reform, and how the teacher culture and rewards systems will significantly affect changes or the lack of changes in the curriculum and classroom. At colleges and universities, the discussions have been driven by the renewed national interest in improving K–12 mathematics and science education and, more recently, the undergraduate curriculum. As recent federal support patterns have shifted toward greater emphases on educational and human resources issues, even the major research institutions are engaged in discussing reform. At the school level, the increased national and state expectations for mathematics and science education, accompanied by a proliferation of new in-service and pre-service programs, have fueled the debate. The State Systemic Initiatives recognize the key component of changing the culture and rewards of the schools and system. In this brief talk, we can only

[1] Former President Bush

focus on one or two aspects. In line with the goal of the conference, I will discuss some directions for school and college faculty which would play a major role in changing the future culture.

Turning first to the schools, a key ingredient to the implementation of innovative or reformed mathematics curricula is a better understanding by the teacher of the content and a better appreciation of its role and connections to other areas in mathematics. This can be achieved only if:

- we have a K–12 mathematics teaching staff with a discipline-based approach to education.

On first glance, many might argue that this is already happening since many pre-service education students are now getting discipline degrees. While that may provide them with a discipline background, it does *not* follow that their approach to mathematics education will be discipline-based. In reality, teacher education is primarily based on cognitive psychology. Also, virtually no school or district has a discipline-based approach to mathematics education, due to the background and attitudes of many of the teachers as well as the school administrators.

While it would take many years to change each K–12 teacher to meet this objective, changing the system's viewpoint about its teaching staff could be achieved in the near future. Once this point of view is adopted, many aspects of systemic reform and teacher culture could be improved, at least for those teachers who believe in this approach. In an article in the *American Educator*, **[2]**, David Cohen describes his observations about a California mathematics teacher recognized by the schools for involvement in reforming her mathematics instruction. His personal and detailed analysis revealed a rather flawed approach in the teacher's instruction, much of it due to her lack of mathematical knowledge. Worse yet, the system's viewpoint toward discipline knowledge and the shallow approach to supporting experienced teachers precluded any help. With the viewpoint above, more district support could be provided, such as time to think, discuss and read mathematics, and preparing curricula with like-minded colleagues.

There are many benefits for the teachers who view the world of education through discipline-based glasses. One striking feature of many European and Asian secondary teachers is their view of themselves as mathematicians and scientists. This perception would likely lead to better relations and connections with university and college faculty and industrial scientists, with teachers viewing college faculty as colleagues, not as trainers or sources of funding. Many summer and academic year programs, such as those offered through the Geometry Center at Minnesota and the three regional geometry institutes, are more valuable to teachers who can first absorb the mathematics and then consider its use in education, rather than immediately expecting direct application to their own classroom setting. As Stevenson and Stigler have noted in their studies of Asian teachers **[12]**, the important feature of teacher encouragement of widespread student participation during mathematics instruction seems to be directly linked to the teacher's security about the depth of his/her own mathematical training. In these same directions the programs and attitudes of schools towards their brightest mathematics students is frequently linked to the teacher's views and interest in mathematics. In a *Focus* article describing a visit of 25 mathematically gifted high school students to Russia [**8**], Mark Saul, the president of the American Regional Mathematics League (ARML), notes the impact of the Russian culture and passion for mathematics on sustaining their amazing output of mathematical talent in chaotic political situations: "Russian mathematicians seem to develop a passion for their subject quite early, retain it for their entire professional career, and are anxious to impart it to a new generation. These cultural elements, particularly the last, are much less developed in American mathematics education." Finally, the discipline-based approach would ensure that as disciplines change (and we can be sure that they will change more rapidly in the future), there will be better chances to bring the new ideas into the curriculum.

Let us briefly look now at the undergraduate curriculum. Here the situation is somewhat different. There, an understanding of research about teaching, learning, and other educational issues is frequently either poorly formulated or ignored by the faculty. At a recent meeting at the National Academy of Science to discuss the Academy's possible contribution to undergraduate science, mathematics, and technology education, it was noted, "If undergraduate reform is to be successful, then attitudinal and cultural changes are required in the teaching faculties. They must be convinced that how a course is taught is as important, and as hard to do well, as content; that effective undergraduate teaching is rewarding in itself and pays off. In short, a set of rewards that motivate faculty to pay attention to undergraduate teaching is necessary. This must happen at the administration level; depending on the individual departments alone won't be effective."

We can effect these types of changes only if:

- We have a collegiate or university mathematics faculty with an educational-based approach to the discipline.

This view of orienting the faculty's undergraduate, graduate, and even research endeavors towards an educational objective is already being discussed and tried in several ways. The programs at the Geometry Center and the three regional geometry institutes are models of vertical integration, where K–12 teachers and students, undergraduates, college teachers, graduate students, post-doctoral fellows, and research mathematicians all engage in various programs linked by an educational mission. At a March 1992 Mathematicians and Education Reform (MER) Network conference, Judy Sunley, the Division Director of Mathematical Sciences at NSF, noted:

- It is a disciplinary responsibility to respond to educational concerns, not just the responsibility of education;
- The mathematical sciences are at a crossroads. We need to respond

> to a larger educational mission, or we may go the path of the arts in federal funding.

It is important to understand that this educational-based approach does not diminish the values of research and scholarship. Rather, it redirects the faculty's goals towards the communication of this knowledge to others, and the education of future generations.

There are many other reasons why our faculty would benefit from such an approach. A 1991 study on university-industry innovations and technology transfer, reported in the University of Minnesota Research Report, concludes that [7, p. 3]: "Academe's primary role—education and provider of talent—is its greatest contribution to the process of innovation" and not in direct innovation. The quotation above by Mark Saul highlights the involvement of Russian university faculty in education. With an educational-based approach, it would be far easier for our faculty to engage in K–12 educational activities and nurture mathematical talent at all levels. Finally, the keys to understanding and implementing curricular reform in our undergraduate mathematics program are the faculty's beliefs and educational orientation. After several years of NSF funding for a major calculus initiative, a few promising models have emerged, together with a lot of questions. Basic questions such as the purpose and goals of calculus still remain under debate. In pre-service teacher education, the faculty are increasingly being asked to teach these courses in ways consistent with the recommendations of the NCTM standards. Most faculty are unaware of the NCTM recommendations. Even if they are aware, very few are pedagogically prepared to effectively use the teaching techniques which we want teachers to use in their classrooms. The massive amount of faculty development necessary to implement teaching to the recommendation in the *Standards* could only be justified under an educational-based approach to the discipline.

Is it possible for the large research institutions, which educate many students, to adapt this educational perspective? This will clearly be a major challenge and take years to achieve. However, support for adopting this viewpoint continues to grow. James J. Duderstadt, President of the University of Michigan and Chair of the National Science Board, said in a 1990 address to the Sigma Xi Scientific Research Society [**9**], "...over the long term, it is clear that we must reform the educational system, that is, completely rebuild the pipeline to respond to the changing world in which we live. In our colleges and universities, it is time to think about improving what we teach, whom we teach, and how we teach." The wedding of education and research provides an excellent vehicle to achieve reform.

C. Mathematical Curricula for the Future

In speculating about curricular changes in mathematics, it is important to keep in mind some historical principles:

- Most of our mathematics is based on historical traditions that are widely accepted within the teaching community;
- There are good reasons to believe that the fine details of mathematics, computation (i.e., arithmetic) and calculation (i.e., algebraic and analytic manipulation), are indeed important in learning to use mathematics at the expert level. Most of mathematics education today is learning the fine details, and not concerned with mathematics literacy;
- There is a great deal of inertia in the system. Change comes slowly and only after a great deal of widespread acceptance;
- What is taught in the classroom can be very different from what is intended to be taught.

With this background, the widespread acceptance in the mathematics K–12 and collegiate community of the NCTM Curriculum and Evaluation *Standards* is indeed impressive. Virtually every school and the vast majority of teachers accept alignment of their mathematics curriculum with the goals of the *Standards*. Hence, the content directions spelled out in the *Standards* will almost certainly be the dominant influence in the K–12 mathematics curriculum for many years. The *Standards* will serve as the major framework for K–12 curriculum discussion and modification and the benchmark for more radical curricular thrusts. In fact, the widespread acceptance of the curricular framework of the *Standards* even allows discussion of a national curriculum—an unthinkable topic to bring up just a few years ago. People like Stevenson and Stigler who take an international view of mathematics education are now more actively discussing the possibility of a national curriculum.

What may be most valuable about the *Standards* are the goals for what we want to teach and assess in mathematics. Indeed, a potential major influence of the *Standards* on the collegiate curriculum may be the growing belief that the appropriate collegiate standard for teaching the mathematics content courses to pre-service mathematics teachers must be close alignment with the teaching and instructional objectives in the *Standards*. It could very well be the case that some of the new curriculum to be developed for the mathematics courses in the pre-service program might even spread to the entire undergraduate mathematics curricula.

The specific content suggestions of the *Standards* are rather conservative. The curriculum is solid and important, but not very reflective of some newer ideas or different directions. Moreover, the *Standards* do not address the desirability of learning or visualizing elementary approaches to advanced content made possible by new uses of technology. Accepting the *Standards* as the mainstream basis for the future curricula, I will conclude by discussing a few examples of alternate approaches as models for future curricula.

1. Broad thematic-based content. Instead of the traditional and somewhat artificial dissections of the K–12 curriculum into separated topics, we could look at courses based on broader themes. In the NRC book, *On the Shoulder of Giants*, five essays are developed on central strands or themes in mathematics: dimension, quantity, uncertainty, shape, and change. Other themes, such as symmetry, could also be used. These strands could form the basis of courses that would continue to be addressed at different levels throughout the K–12 mathematics curriculum. In the American Association for the Advancement of Science Project 2061 report, a model for problem solving using statistical principles—prediction, model, uncertainty (careful examination), and optimization (finding maximum and minimum)—is developed. This model could be used in different types of curricula. One point is clear. In order to pursue thematic-based curricula, it would certainly require teachers educated in and practicing a discipline-based approach towards mathematics education and a very enlightened view of what is valuable in mathematics. The sciences may be our best allies in this approach. The curriculum suggested in Project 2061 removes rigid boundaries between traditional disciplines and develops underlying themes and the 'big picture'. If the forthcoming science standards now being developed by the National Research Council take the same approach, thematic-based curricula in mathematics may soon follow.

2. New views of traditional courses and advanced topics. The notion of geometry and what teachers might teach as envisioned and demonstrated at the Geometry Center in Minnesota is very different from what happens in most classrooms. Using both new variants on traditional ideas (peeling vegetables to observe curvature, sewing Klein bottles) and powerful visualization from the latest computer software, the Center is trying to develop modules to be included in the standard secondary geometry curriculum. The topics should include hyperbolic geometry, curvature, planar tilings, and exploration on surfaces in 3-space. A highly innovative course, *Geometry and the Imagination*, was developed and taught in 1990 at Princeton and in 1991 at the Center. The notes have been published and materials, videos, and modules are now in development. Equally imaginative materials are being developed using technology and visualization in other areas, such as fractals and chaotic behavior of iterated maps. These topics, because of their currency to applications, visual beauty, and close connection to deep mathematics, have a real appeal for many students. They may provide opportunities for revitalized curricula that would help to attract more students into studying and enjoying mathematics.

There are opportunities even within the more traditional topics of the mathematics curriculum. The familiar function $f(x) = \sin x$ looks dramatically different using computer printouts in which a lot of discrete data from the graph is compressed over a large scale (see Figure 1). It challenges our image of the familiar periodic function and even why it is the graph of a function.

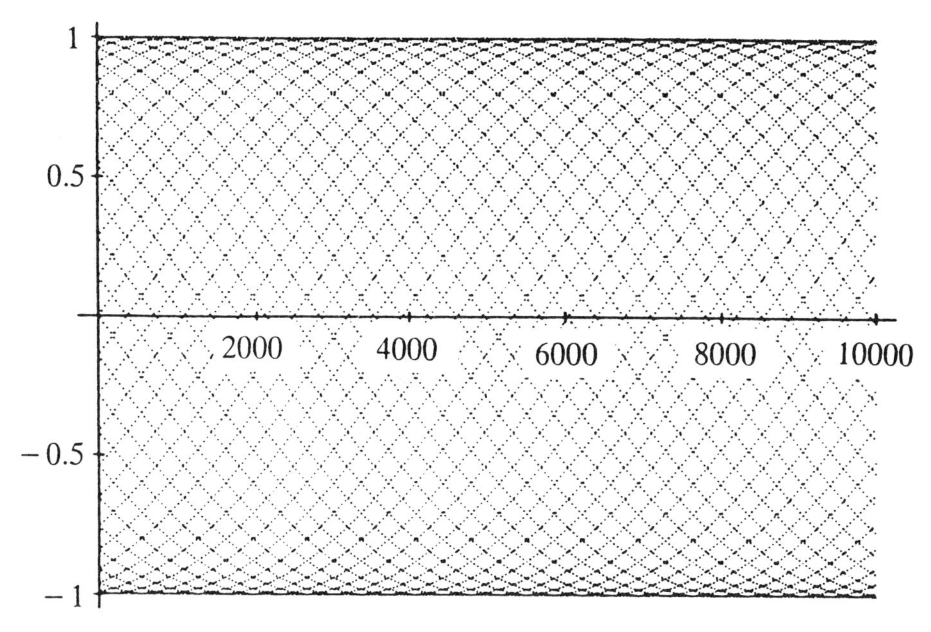

FIGURE 1. 10,000 points of $\sin n$, $n = 1, 2, 3, \cdots$. What's the explanation of the many periodic curves?

This creates new opportunities to discuss functions and their graphs, and really begin to better understand deep concepts such as domains, range, and graphs of functions. The latest graphing calculators, such as the TI-81, enables large numbers of students to continue to pursue similar discussions, and hence make a really central notion of mathematics—functions—become alive again.

Finally, we can frequently do quite a bit of curriculum innovation by just letting our mathematical imagination take a fresh look. In a model calculus curriculum being developed at Harvard University, they again focus on the notion of graphs and functions and expect the student to associate a graph with a verbal description of a seemingly nonmathematical situation (see Figure 2 on next page).

Many mathematicians get very excited about these approaches to functions because they believe that this is the real content of understanding functions. Moreover, they provide nice, geometric, noncomputational ways to learn about functions and graphs, exactly the kind of approaches that the *Standards* recommend. If K–12 teachers viewed functions and other mathematical ideas through the discipline-based glasses that we discussed, then probably many topics besides functions would also get a fresh look and more mathematically exciting approaches might be developed.

So now we have completed my overview of the many different aspects necessary to really change the K–16 mathematics curricula and keep the process of systemic reform of mathematics education moving forward. It is quite a

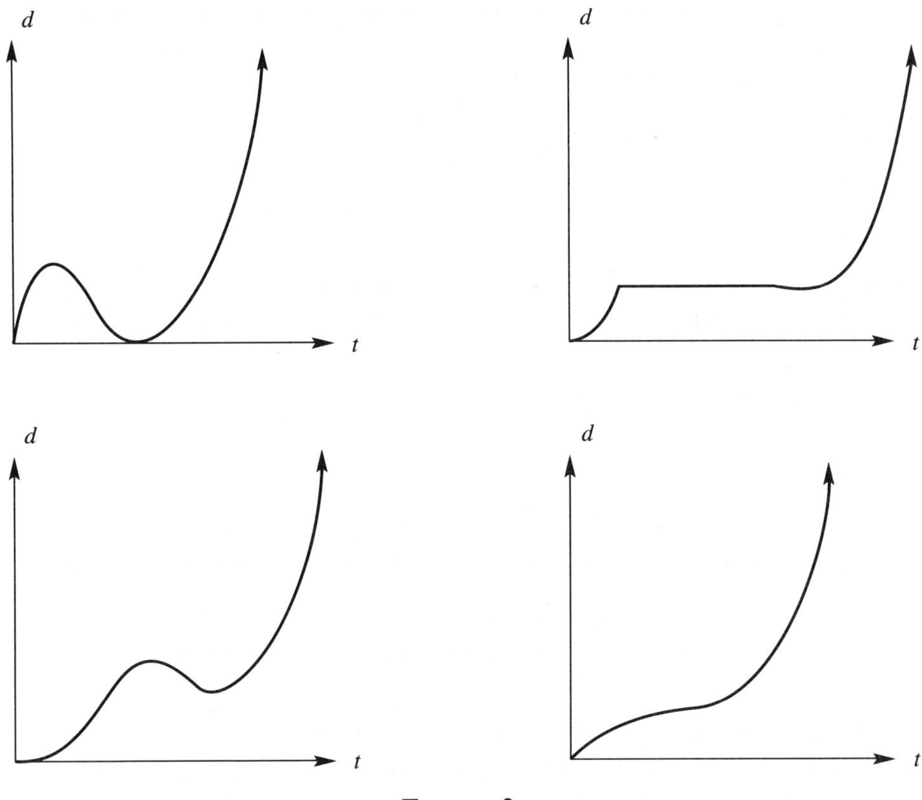

FIGURE 2

daunting task, and will require substantial effort from many individuals and some very challenging modifications of our values and institutions. Your deliberations at this conference are important steps in reaching these goals.

REFERENCES

1. American Association for the Advancement of Science, *Project 2061: Science for all Americans*, American Association for the Advancement of Science, Washington, D.C., 1989.

2. David Cohen, *Revolution in one classroom*, American Educator, American Federation of Teachers, Fall 1991.

3. Harvey Keynes, *Equity and excellence in the University of Minnesota Talented Youth Mathematics Program* (*UMTYMP*), Amer. Math. Soc./CBMS Issues in Mathematics Education, vol. 2, 1991, 85–96.

4. National Center for Educational Statistics, *The state of mathematics achievement: NAEP's 1990 Assessment of the Nation*, U.S. Dep. of Education, Washington, D.C., 1991.

5. National Council of Teachers of Mathematics, *Curriculum and evaluation standards for school mathematics*, National Council of Teachers of Mathematics, Reston, VA, 1989.

6. National Research Council, *Everybody counts: A report to the nation on the future of mathematics education*, National Academy Press, Washington, D.C., 1989.

7. Research Review, *The primary role of universities*, Newsletter of the Office of Research, University of Minnesota, January, 1992.

8. Mark Saul, *Love among the ruins*, Focus, MAA, vol. 12, February, 1992.

9. Patricia Shure, Donald Lewis, et. al., *Women in mathematics and physics: Inhibitors and enhancers*, University of Michigan, Ann Arbor, MI, December, 1992 (preprint).

10. Lynn A. Steen, (ed.). *On the shoulders of giants: New approaches to numeracy*, National Academy Press, Washington, D.C., 1990.

11. Lynn A. Steen, *Will everybody ever count?* Third UCSMP International Conference on Mathematics Education, University of Chicago, Chicago, IL, October, 1991.

12. Harold W. Stevenson and James W. Stigler, *How Asian teachers polish each lesson to perfection*, AMP Line, Newsletter of the American Mathematics Project, Winter 1992.

13. TIMS, *Extrapolator*, Newsletter of the TIMS Project, University of Illinois at Chicago, Chicago, IL, Fall-Winter, 1991–92.

14. U.S. Department of Education, *A study of characteristics of stayers, movers, and leavers, 1988–89*, U.S. Department of Education, Washington, D.C., 1991.

15. Zalman Usiskin, *Teaching and testing in the 1990s*, UCSMP Newsletter No. 10, 1992, 7–11.

16. Who's Who Among American High School Students, *Twenty-second annual survey of high achievers*, Educational Communications, Lake Forest, Illinois, 1991.

DEPARTMENT OF MATHEMATICS, UNIVERSITY OF MINNESOTA, MINNEAPOLIS, MN 55455
E-mail address: keynes@math.umn.edu

Belmont University Library